A Thousand Answers

TO

Beekeeping Questions

BY

DR. C. C. MILLER

As answered by him in the columns of the
American Bee Journal

COMPILED BY

MAURICE G. DADANT

PUBLISHED BY
AMERICAN BEE JOURNAL
Hamilton, Illinois
1917

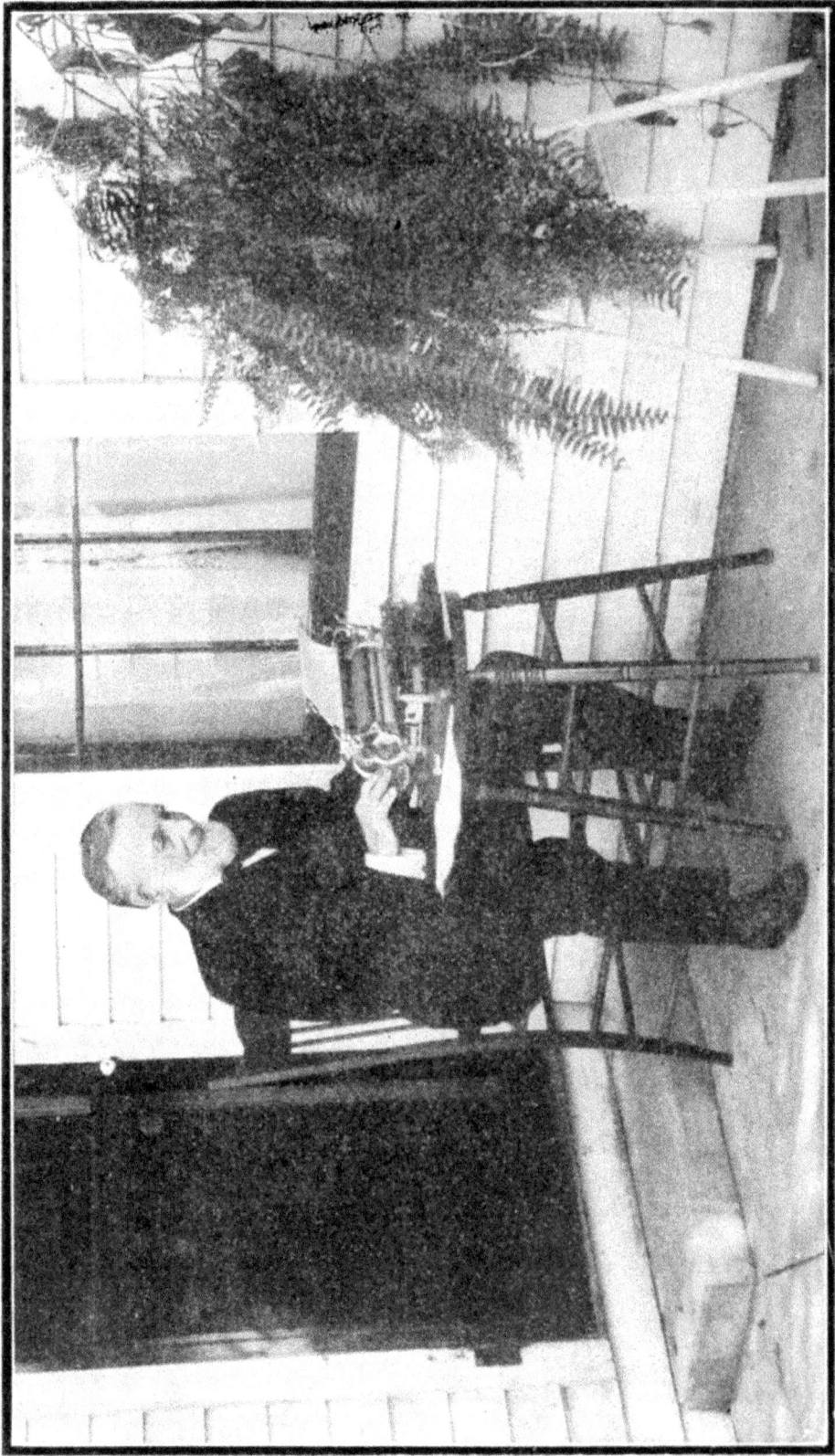

Dr. C. C. Miller at his typewriter answering questions for his regular department in the American Bee Journal.

A Thousand Answers

TO

Beekeeping Questions

BY
DR. C. C. MILLER

As answered by him in the columns of the
American Bee Journal

COMPILED BY
MAURICE G. DADANT

PUBLISHED BY
AMERICAN BEE JOURNAL
Hamilton, Illinois
1917

Northern Bee Books

A Thousand Answers to Beekeeping Questions © Maurice G Dadant

First published by the American Bee Journal, Hamilton, Illinois, 1917

This edition published by Northern Bee Books 2013
Scout Bottom Farm
Mytholmroyd
Hebden Bridge
HX7 5JS (UK)

ISBN 978-1-908904-37-9

D&P Design and Print
Worcestershire

Printed by Lightning Source, UK

DR. C. C. MILLER

Personal Reminiscences of the Edit-or, who had Known him for Nearly Forty Years

By E. R. Root

A GREAT voice has been still-ed; but those bright and bree-zy sayings from the Sage of Ma-rengo, always labeled with smiles, will live after. Such a life can not die; but all that is earthly of Dr. C. C. Miller passed away on Sept. 4.

When he was obliged to give up his de-partment of Stray Straws some months ago, on account of a severe sickness and his advanced age, there came a feeling over me that I must see him once more before he passed from the scenes of earth, feel his handshake, and see that face so beam-ing with smiles.

As I was scheduled to be present at a Chautauqua held at Madison, Wis., on Aug. 16 to 20, I decided that on my return I would pay Dr. Miller a visit between trains, during which I would take some more pic-tures of him; for I felt sure that the bee-keeping world would want to see him in his ninetieth year. On arriving at the Chau-tauqua I told Dr. E. F. Phillips that I pur-posed to go and see the man who wrote Stray Straws, and asked him if it would not be possible for him and Mr. Demuth to go along with me. Precisely that thought was in the minds of both of these men, and we were not long in making up a little party to motor from Madison to Marengo. This party was made up of Dr. E. F. Phil-lips, Geo. S. Demuth, H. F. Wilson, and the writer.

We had expected to see Dr. Miller show-ing his age, and that the once virile face and form would be infirm with years; but we were agreeably surprised to see ap-parently the same man with the same vigor of body and mind that I had seen 35 years ago. He seemed to be at his very best, and members of our party all agreed that his mind was as alert and keen as ever. I think the average person would have said that he was not over 70, and probably along in his 60's.

But that wonderful smile that betokened the happy nature within must have camou-flaged whatever of bodily infirmity there might have been. And surely there was some, because he died just two weeks to a day after our visit. I said, "Doctor, I'd give 20 cents for a picture or two of you;" and instantly he came back with a laugh, saying: "Beg pardon. I'll have to charge you 35 cents this year." At this the camera clicked, and the result is shown on the cover page of this Gleanings.

I had told him I had come to convey the best wishes of my dear old father, and it gave me pleasure to tell the Doctor of the joy that his letter (published on page 624 of this issue) gave to A. I. Root. I further added that father wanted to pay him a visit, and hoped that he might yet do so. I shall never forget how that smile seemed to fade a little, and then how it came back with its wonted sweetness in these words:

"I should dearly love to see your father again, for he and I are about the only ones left of the old group. But tell him he must come soon, as sometimes I think I have not many days to live. If I do not see him on this side, I surely shall on the other side."

As he said that, the camera clicked again.

I took of him that day some two or three dozen pictures, and in future issues I hope to show more of them, as it will take a good many pictures to show the many sides of this wonderful man.

At this time I wish to give a few per-sonal reminiscences, beginning with the time I made my initial bow to the beekeep-ers in the early 80's, or about the time that my father's health broke down and his editorial mantle was thrown on me. It was at that time I needed the help both of my father and of Dr. Miller. I needed Dr. Miller because father's enthusiasm was fast drifting toward gardening and green-houses and other like pursuits; but not so with Dr. Miller. I remember how, after I had come back from a trip among the bee-keepers of New York, I thought I had gathered some new ideas. I had felt that the hives and appliances we were then mak-ing would have to be modified to fit com-mercial beekeeping not only on a large but on a small scale. For example, I became satisfied that father's beveled edge of the Simplicity hive and the metal-corner frame would have to give way to the square edge and the all-wood frames that were then coming into use. Naturally father was conservative. To settle the question we agreed to call in Dr. Miller. To make a long story short, the dovetailed hive was first launched on the market in 1889. Slow-ly it crowded out all its competitors until it is now the standard of all the hive-manu-facturers in the country.

In those early days I needed Dr. Miller's help again in launching the thick-top frame. In fact, Dr. Miller had proposed it to me himself, saying that it was a great step forward, as it would effectually rid the hive of burr-comb, and it did. With Dr. Miller's support I put this in the catalog, and now the thick-top frame is in almost universal use among beekeepers.

About this time, also, I advocated self-spacing frames, and especially Hoffman frames. Here Dr. Miller expressed his doubts. He later came to see the value of the self-spacing feature in the form of nails as spacers; but he never really accepted the Hoffman frame, now in general use.

When the Porter bee-escape was first put on the market, Dr. Miller was again doubtful; but experience soon showed to him that it was a great invention for clearing bees from the supers.

During the time that the divisible-brood-chamber hives were being exploited in the 80's, I remember that Dr. Miller was skeptical, saying he doubted if the principle were correct. Mr. Heddon, Mr. Hutchinson, Mr. Taylor, and scores of other leading beekeepers at that time advocated the principle of handling hives rather than frames, and it certainly did look good; but Dr. Miller said to me privately: "You will do well, Ernest, not to push it," and we never did. The years that have gone by since then have proved that it was a step backward. It is not surprising, in the light of our present knowledge, that those who advocated and used this divisible-brood-chamber system of honey production had so many failures that they began to think that the seasons were to blame. They never seemed to think it could be the hive with its little force of bees.

Experience during the last 15 or 18 years has shown that, instead of dividing up a brood-nest, we should double them up and make strong, populous colonies. Dr. Miller, some 20 years ago, supported my contention that a good queen needs at least two eight-frame hive-bodies for breeding purposes. I advocated at the time a double brood-nest—not a brood-nest split in the middle or in halves, as advocated by Mr. Heddon. Just at the beginning of the honey flow, when running for comb honey, the plan was to reduce the breeding room to one-chamber, forcing all the bees into the supers. Dr. Miller made this a practice for years. He never had any trouble about getting bees into the supers, for the reason that he had the hive so "chock full," as he said, that they simply had to go into the sections. He was getting crops of honey right along when the users of divisible-brood-chambers were complaining of poor seasons one after another. Likewise, Dr. Miller always supported the Dadants in their advocacy of large brood-nests; or, as he used them, a double brood-nest of nearly the same capacity.

Dr. Miller was almost the first one to see that horizontal wiring, while it made beautiful flat combs, as smooth as boards, resulted in the foundation stretching near the top-bar, thus making the cells, when drawn out, too large for the queen to lay in. We therefore find him, some 20 years ago, advocating wood splints. While, possibly, this is not the best means to prevent stretching, it is a good one, and goes to show how Dr. Miller was looking forward, and how he was ahead of the times.

The grand old man of beedom never claimed to be an inventor. He never claimed he had any secrets, for he had none. His great service to the bee world was in discovering practical methods for producing more and better honey with the appliances that the beekeeper has. He never was in favor of throwing away old hives or apparatus as was Mr. Heddon; and therefore one would never find anything in the Doctor's apiary but standard hives, standard Langstroth frames, and standard equipment sold by every supply dealer in the country. While he did not invent, he did pick out of the mass of crudities inventions that he approved.

I have just said that Dr. Miller did not pretend to be an inventor; but there are some things that bear his name—the Miller feeder, for instance; but he was generous enough to say that Mr. Warner improved it so that it was better than his own feeder. An introducing-cage also bears his name. This was not exactly an invention, but it was improved so that it is really a practical introducing-cage, one that is used very largely by queen-breeders.

There is hardly a standard article sold by manufacturers, now accepted by the beekeeping public today, that was not passed upon by Dr. Miller before it went on the market. For example, the eight and ten frame dovetailed hive was submitted to Dr. Miller at Medina before being introduced to the public. In fact, neither A. I. Root nor the other members of our organization thought it best to put anything on the market unless it had Dr. Miller's approval. In the same way brood-frames, self-spacing frames, bee-escapes, and introducing-cages were passed before the critical eyes of Dr. Miller. If he pronounced them good they went to the public. The fact that these things have been in use for 20 and even 30 years by practical beekeepers all over the United States shows how nearly Dr. Miller was right.

Let us now look a little further and see what Dr. Miller did in making bee-culture saner and safer. Perhaps the biggest thing he ever did was to show to the world the real nature of European foul brood. He blazed the way in perfecting a new cure for that disease—a cure that is accepted today. E. W. Alexander furnished the basis for the treatment, and S. D. House, Camillus, N. Y., showed that the period of queenlessness could be reduced. He also showed that a resistant stock of Italians would go a long way in curing the disease and keeping it out of the apiary. But the ideas advanced above by Alexander and House were so revolutionary that there were but very few who took any stock in them. Only too well do I remember how I was criticised for publishing these "false" doctrines. But it was not until Dr. Miller had tried them out and had proved that they were along right lines that the beekeeping world began to take notice. The good Doctor went further than either Alexander or House in showing the true nature of the disease, and, possibly, how it spreads. When, therefore, Dr. Miller introduced these new methods of treatment the whole of beedom turned right about face. Later work by Dr. Phillips and his assistants proved the soundness of Dr. Miller's views.

Dr. Miller, later on, developed, if he did not invent, a plan for uniting bees with a sheet of newspaper. The plan is very simple and effective. He moved the weaker of the two colonies to be united and placed it on top of the stronger one. Between the two stories was placed a sheet of newspaper (with or without a small hole punched in it). The bees would gradually unite thru this paper; and because the uniting was so gradual there would be no fighting and less returning of the moved bees to their old stand.

This little sketch would be incomplete, were I not to refer to a very predominant and dominant characteristic in Dr. Miller— that temperament or quality in his nature that makes the world delightful and everything lovely—so much so that it showed out not only in his face but in his writings. I think some of the happiest times of my life have been spent in Dr. Miller's home. Not only did he carry optimism thru the printed page, but we found it at the breakfast-table and all thru the day without a let-up. He went further. His conversation was one ripple of merriment thruout. He never ridiculed, but he could see the funny things of life, and sometimes I have come away from his table sore from laughter. He had the habit of taking one by conversational surprise, and would have him holding his sides almost before he knew it.

I said to him 30 years ago: "Doctor, I wish there were some way by which you might reproduce those breezy remarks you make at conventions and in your home— those little sidelines that are so helpful and yet seem like a drink of cold water on a hot day. Is it not possible that you could send Gleanings a page or two of short items of general comment each month? and I would suggest the name 'Kernels of Wheat,' as we already have a department, 'Heads of Grain.'"

He liked the idea; but for a title he suggested that "Stray Straws" would be much more appropriate. That would be more in line with his ability, he said. Our older and younger readers know how well he succeeded in giving us "Stray Straws". They were really kernels of wheat. Dr. Miller's paragraphs of five to a dozen lines were worth whole articles; and almost every one of those paragraphs was replete with smiles.

Years afterward, when I talked about the success of his department he said to me: "Ernest, all the credit belongs to you. You discovered how I might be able to give a little help to beekeepers, and I am certainly glad if I have succeeded."

Years ago at some of the conventions there was more or less strife; and well do I remember that Dr. Miller, in his quiet way, with a smile that was more persua-

sive than a policeman's club, would smooth out all the difficulties leaving a good feeling all around. In this respect he and Prof. Cook were without a peer. I remember one day he came to me, in the history of the National Beekeepers' Association, when there seemed to be a bitter fight on. He said to a group of us: "You have asked me to pour oil on the troubled waters. The job is too big for me, boys. But I will try my best if you will offer a prayer that only good may prevail"—and it did.

This brings me to another important side of Dr. Miller's character—an abiding faith in God. Come what might, with him all was well. There came a time when, thru some mismanagement on the part of others, he lost a considerable part of his savings. With a sweet spirit of resignation he wrote: "I have not lost all. I have my good wife and my sister. I have a few years of vigorous life left to me yet. I have in prospect a good crop of honey. The Lord has always taken care of me, and I am not worried over the future."

Dr. Miller would have been great in any line of work or profession. Had he stayed in music his fame would have gone over the world, I verily believe; and if he had kept on in the practice of medicine he would have advanced the profession materially. Even in the early days he said people did not need medicine so much as they needed common sense in treating their bodies. He gave up the practice of medicine because he said he did not believe much in giving medicine, and because he had to charge for his daily visitations; and, because his patients objected to paying his bills when he had given no medicine, he would go into something that was more congenial to him.

Let me tell you why I think Dr. Miller would have been great in the field of medicine, or, I should say, in healing. In his day medicine was considered as almost the sole reliance, but not so with Dr. Miller. Fifty years ago he believed that hygiene, plenty of water inside and out, rest, and temperance in eating, are far more important than drugs. Our best doctors today would testify that he was fifty years ahead of his time. The modern schools of medicine are advocating less drugs and more hygiene, plenty of good air and water. When Dr. Miller was going thru college he did not know that he could overwork, but soon found that he was burning the candle at both ends. He came out of college a full-fledged graduate with several hundred dollars to the good, but with health broken. All his life he had to be careful what he ate, as a consequence. He was obliged to keep from overeating as well as from overdoing. Many and many a time I have seen him at the table stop short. "I would like to eat that," he would say, but he would rigidly deny himself, and the result was that he kept himself active in mind and body. He was not only a great teacher but a great healer.

PREFACE

IN 1895 there was begun, in the American Bee Journal, a department of "Questions and Answers," with Doctor C. C. Miller in charge, the object being to give information to readers on special subjects, perplexing to the beekeeper, and not specifically covered by the different bee literature.

In the twenty-two years that Doctor Miller has answered these queries of subscribers (he is still conducting this department) almost every subject in beekeeping has been touched. His wide experience, his inimitable style, and the clearness with which he writes have made these answers invaluable.

The present volume is a compilation of a thousand questions, culled out of many thousands and arranged in alphabetical order for convenience. Its object is not to supplant existing text-books on beekeeping, but rather to supplement them.

MAURICE G. DADANT.

DR. MILLER'S THOUSAND ANSWERS

Absconding (See also Desertion, Swarms Leaving.)—Q. I hived a swarm, and the next day it sailed off to parts unknown. What shall I do to prevent such a thing in the future?

A. The most frequent cause of such desertion is heat. A hive unshaded standing out in the boiling sun with a very small entrance and all the rest closed up tight, is a pretty warm place to set up housekeeping, and one can hardly blame the newly settled family for moving out.

The remedy is not difficult to imagine. If possible, let the hive be in a cool, shady place. A temporary shade, and sprinkling with water will serve a good turn. Give plenty of chance for air. Some practice leaving the covers of the hives raised an inch or so for two or three days. Some give two stories to the swarm, taking away the lower story after two or three days. Either of these plans provides to some extent against an overheated dwelling. Some practice giving a frame of brood to the swarm, with the idea that the bees will feel that they cannot afford to abandon so valuable a piece of property.

In any case, if all laying queens are clipped no prime swarm can abscond unless it joins, or is joined, by some other swarm having a queen with whole wings. The queen with clipped wing may be lost, but it is better to lose the queen alone than to lose both queen and swarm.

Swarms may abscond, also, if they are secondary or after-swarms and the queen has not mated. When she goes out for her wedding flight, the swarm may follow her.

Absorbents.—Q. Which is the better way to fix bees for wintering out-of-doors, with a tight-fitting cover on the hive, or with chaff cushions, or some other porous absorbent material?

Is there any way to keep the moisture from the bees, and from condensing in the hives? If so, how?

A. If a plain board cover be directly over the bees the moisture will condense on it and fall on the cluster; but the moisture will not condense so readily on wool, chaff or something of that kind; so that it is preferable to the close-fitting board cover.

Adel Bees.—Q. Is the Adel bee a sort of Carniolan bee, and

can it be kept in an 8-frame hive?   What kind of a cross would
it be?

A.   There is no such race as Adels.   The word "Adel" is a German word which Germans spell "Edel," and the word means noble
or excellent.   So anyone may call his bees Adels, whether they
are black or yellow; only, of course, it will be a misnomer if applied to poor bees.   If I understand it correctly, Adels were a
strain of Italians first, so named by Henry Alley.

Afterswarms.—Q.   My bees swarmed May 31.   I put on a super
that noon, and eleven days later they put off another swarm.
What was the matter with them?   They have not started to build
in the super yet, and the new bees are still bringing in honey in
the bottom.   What is the reason?

A.   It is the usual thing for bees to send out the second swarm
about eight days after the prime swarm, and it may be as much
as sixteen days later.   They may also send out a third, fourth
swarm, or more, and even if they send out only one swarm they
are not likely soon to do anything in the super, if at all.

Q.   Last spring I bought three colonies of bees from one of the
neighbors and they all have crooked combs in the brood-chamber.
He did not use starters, and they are so crooked that I cannot
take them out of the frames.   These same colonies have each
swarmed three times.   The first swarms were large.   I hived
them in new 10-frame hives.   The next three swarms were smaller.
I also hived them in 10-frame hives, and the last three were
small.   As I did not want any more bees, I killed the queens in
the last three swarms and put them back in the parent hives.
They did not swarm any more.   As I don't want any more swarms,
how can I prevent them from swarming?

A.   One way of preventing too much increase is to do as you
did in one case, that is to return the swarm as often as one issues.   But that may be more trouble than you like.   Here's an
easy way to prevent afterswarming: When the prime swarm is
hived, set it on the stand of the old colony, setting the old hive
close beside it, facing the same way.   A week later move the old
hive to a new stand 10 feet or more away.   That's all; the bees
will do the rest.   For when the hive is moved to a new stand the
bees will go to the fields just the same as if they had not been
moved, but when they return, instead of going to their own hive
they will return to the old stand and join the swarm.   That will
so weaken the mother colony that all thoughts of swarming will
be given up, especially as no honey will be brought in for a day
or two after the change of place.   If you want to prevent all

swarming that's a more difficult matter. Inform yourself thoroughly by means of such a book as Dadant's Langstroth, and you

FIG. 1. Afterswarms many times are accompanied by several queens and
cluster as above.

will be in a better position to know what plan is best for you.
My book, "Fifty Years Among the Bees," is especially full as to

the matter of hindering swarming. But I must confess that I have not been able to prevent all swarming to my entire satisfaction. It may be some help to say that if you succeed in getting a young queen reared in a colony and get her to laying, that colony is practically certain not to swarm the same season.

Q. In preventing afterswarms, by placing the young swarm on the old stand and taking the old colony to a new place, should all the queen-cells except the ripest one be cut out at once?

A. That's one way. There's a better way. Set the swarm on the old stand, the old hive close beside it, without cutting out any queen-cells, and let stand for a week. Then move the old hive to a new stand, and the bees will do the rest. You see, when the old hive is moved at that time all the field-bees will leave it and join the swarm. That will weaken the old colony, and added to that is the fact that no honey will be coming in, so the bees will conclude they cannot afford to swarm, and all the extra queen-cells will be killed without your opening the hive.

Q. Can an afterswarm be returned to the parent hive? If so, how shall I proceed?

A. The easiest thing in the world. Just dump the swarm down in front of the hive and let them run in. It was the old-fashioned way of treating afterswarms, and there's no better way, if you don't mind the trouble. Just return the bees every time they swarm out, and when all the queens have emerged there will be only one left, and there will be no more swarming. Indeed, you may carry the plan still farther, returning the prime swarm and all the afterswarms. That will give you no increase, but the largest yield of honey, especially if your harvest is early.

Hiving the swarm in an empty box and returning it to the parent colony the next day is still better, as the swarming excitement is over.

Q. Do afterswarms come out only when the old hive remains on the old stand?

Do they always fail to come out when the old hive is put in a new location?

A. Afterswarms are likely to issue if the old hive is left on the old stand, and are less likely to if the old hive at the time of swarming is removed to a new place; but may issue then. If the swarm is put on the old stand, the old hive close beside it, and then a week later the old hive removed to a new place, you may count quite safely on no afterswarms.

Age of Bees.—Q. What is the average life of a queen, drone, and worker bee?

A. A queen, perhaps 2 years; a worker, 6 weeks in the working season; a drone, until the workers drive it out.

Q. I have heard a great many say, bees live only 30 days. What do you think about it?

A. Worker-bees live several months if born late in the season; for they live over winter and until new ones are ready to take their place in the spring. Those that are born after the busy season begins in the summer, live 5 or 6 weeks.

Albinos.—Q. In my bee-book I did not find anything stating the difference between an albino and another race of bees. Is there anything peculiar about albinos?

A. Albinos among bees are somewhat like albinos of the human race or other animals; there is a deficiency of coloring pigment. This is accompanied by weakness in other respects; although some have reported albino bees that were good. I have seen nothing about albinos for several years, and don't know where you could find them.

Alfalfa.—Q. Does alfalfa yield honey in the east? I have seen it stated several times in prominent farm journals that alfalfa gives no nectar in the east.

A. I think the rule is that east of the Mississippi River alfalfa never yields any nectar to speak of. Alfalfa grows finely on my place, and occasionally I have seen a few bees on the blossoms, but never to amount to anything, and I think this is generally so east of the Mississippi. Seems to me, however, that a more favorable report has been made by some one in Wisconsin or New York.

Q. What is the flavor of alfalfa honey?

A. Alfalfa honey is of very mild flavor, milder than clover.

Alley Method.—Q. Where could I get Alley's book on queen-rearing?

A. It is out of print, but you will find the Alley queen-rearing method in "The Hive and Honey Bee," latest edition.

Ants.—Q. I have six colonies of bees. The smaller ones are bothered with large, black ants. Is there any way of stopping them?

A. Ants annoy the bee-keeper rather than the bees. It is decidedly annoying to have them crawling over the hands and biting. Yet it may be well to add that there are ants and ants. Go far enough South and you may find ants that will destroy a colony sometimes in short order. Even in the North there is a kind to be dreaded. You say yours are "large black ants." Most likely that means ants that are a quarter of an inch or so in length, which are large in comparison with little red ants. But if you have the big wood ants that are three-quarters of an inch long, then that's another story. I've had no little trouble with them and they are hard to combat. They get between the bottom-board and board on which it rests, and honey-comb the bottom-board. Sometimes there will be merely a shell left, so that you will hardly notice anything wrong, yet a little touch when hauling bees might break through a hole to let the bees out. Carbolic acid may do something toward driving them away. You may also poison them. Take two pieces of section, or, perhaps, better still, two thin boards 4 inches square, or larger, fasten upon each end of one of them a cleat one-eighth inch thick, and lay or fasten the other on it, thus leaving a space of one-eighth inch between the two boards. Mix arsenic in honey and put between the boards. The bees cannot get into so small a space, but the ants can. Or, put poison in a box covered with wire-cloth that will let the ants in but keep the bees out.

Q. How can I rid my apiary of red ants? They build their hills near and sometimes directly under the hives and crawl into the hives and kill the bees.

A. Have four feet to the hive, each foot standing in a vessel of oil or water. Find the nest of the ants, with crowbar make a hole in the nest and pour in carbon disulfide. Have no fire near, as the disulfide is explosive. Gasoline will also answer pretty well.

Apiary.—Q. How many colonies of bees can be kept in one apiary?

A. That depends upon the pasturage within a mile or two. In most places not more than 75 or 100.

Q. My bees have at least 300 acres of clover and alfalfa within two miles. How many colonies can I pasture to be safe?

A. That's one of the very hard things to say. You don't say whether red or white clover. If you mean red, it probably doesn't count for much, while white clover counts heavily in good years, although some years it blooms a plenty and yet yields no nectar. If you had said 300 acres of white clover, meaning 300 acres solidly occupied with white clover, I should guess 200 colonies might get good picking. Alfalfa varies more. If it is all used for raising seed, then it probably counts as much as white clover. If used for hay, it counts for less, and may count for nothing, depending upon the times when the hay is cut. If always cut just before it blooms, then it counts for nothing; if cut when

Fig. 2. A good location, a south-east slope with windbreak of natural shrubbery.

in full bloom, it may count perhaps on being enough for 100 colonies. You will easily see that, as you state it, the whole thing is a varying problem. It may be mostly white clover, or it may be mostly alfalfa, and the alfalfa may be treated so differently as to make a big difference in the amount of nectar got from it.

Q. Is a lawn sloping to the north a good location for bees? The entrances to face the north, and no shade?

A. You will probably find that it will not make very much difference whether the slope and the aspect are toward the north or south during most of the year. Sometimes your north slope will be the better one, and sometimes the south. In cool days the southern exposure will generally be better, and in the hottest days the northern. In winter there will be days when soft snow is on the ground and the sun shining brightly to entice the bees out to a chilly tomb, and on such days the northern aspect will be better. There will be other days in winter when the weather and all conditions are favorable for a cleansing flight, and then the southern slope will be better. That cleansing flight is a matter of such importance that on the whole it is better to have the southern slope for wintering. This refers, of course, to locations far enough north to make a winter flight an infrequent occurrence. If your bees are wintered in the cellar, it will probably be a toss up which way is better.

Apifuge.—Q. I read in my bee-book about apifuge. What is it? Will it really keep bees from stinging?

A. Apifuge is the name of some combination of drugs, which combination is not made public, and is made, advertised and sold in England. I don't remember its being advertised or used on this side. It probably helps to prevent stings. I have seen it claimed that oil of wintergreen rubbed on the hands would prevent stinging.

Associations, Bee.—Q. Where is my nearest bee-association, and what are the annual dues?

A. The secretaries of associations change nearly every year. Write the publishers of your bee journal for information.

Baits.—Q. What do you mean by baiting to get the bees to work? Do you put in sections partly filled with honey?

A. Sections that are only partly filled are emptied of their honey by the bees in the fall, and the next year one or more of these are put into the first super to start the bees. Such sections are called bait-sections, or baits.

Q. In putting bait-sections, or sections partly filled with comb, into supers when you put them on at the beginning of the season, wouldn't the super be filled better, that is, wouldn't all of the sections be more likely to be completed at the same time, if the bait-sections were put at the outside of the super? Wouldn't it be just as effective in getting the bees to go up and begin storing honey in the super? Or, one might have one bait-section in the middle and the rest on the outside.

A. Your views are all right. Bees will start soonest on a central bait; but if more than one in a super, put them in the corners, or at least outside.

Baits for Swarms.—Q. A neighbor places common boxes up in trees and catches stray swarms. Is there anything a person can put in a hive that will bait a swarm to the box placed in a tree?

A. Yes, you can put brood-combs in it. If the combs have been used but are still sweet and clean the bees will like them better than any empty hive.

Banats.—Q. Is the Banat bee a new race of bees brought from some other country, or is it just a cross with some of our native bees? Would they be hardy enough for Minnesota?

A. It is counted a separate race. I know very little about them, but I suppose they are equally as hardy as Italians, and perhaps as good workers.

Barrels.—Q. Where can I obtain barrels for extracted honey?

A. Second-hand alcohol or syrup barrels are best and could probably be obtained from drug stores, groceries and wholesale medicine firms.

Basswood.—Q. (a) I have just ordered some basswood trees. How close can I plant them together?

(b) Will they grow well in this climate; that is, hot and dry in the summer-time, subject to strong winds in winter, no snow, and temperature never falling very low?

(c) How long will it be before they yield nectar to amount to anything?

(d) How much water do they need when growing? (California.)

A. (a) When they get to be large trees, 20 to 25 feet is close enough. It is not a bad plan to plant only half as far apart as you want the trees finally; then when half grown, to cut out three-fourths of them. The danger is that you will be too tender-hearted to cut them at the right time; but you will not have so much nectar from the large trees that are too crowded. You will easily see, however, that up to the time they get half their full growth there would be a gain in nectar by having the larger number of trees.

(b) I don't know. One would think that conditions are all right; yet I don't remember that anyone has reported planting basswoods on a large scale in California.

(c) Not before 8 to 12 years in this locality; but things move faster in your pushing climate.

(d) At a guess, I should say the same amount as crops in general, particularly other trees.

Q. How long does basswood bloom last and what time does it generally begin in Northern Iowa?

A. It probably begins in Northern Iowa not far from the same time as here, somewhere in the first part of July, and lasts 10 days or so.

Beebread.—Q. (a) Can bees live without beebread in the winter-time?

(b) Can they live on beebread a week or two without honey?

A. (a) Yes, but they must have it in the spring, so they can rear brood.

(b) I think not.

Bee-Cellar (See Cellar.)

Bee-Culture—Most Important Thing in.—Q. What do you con-

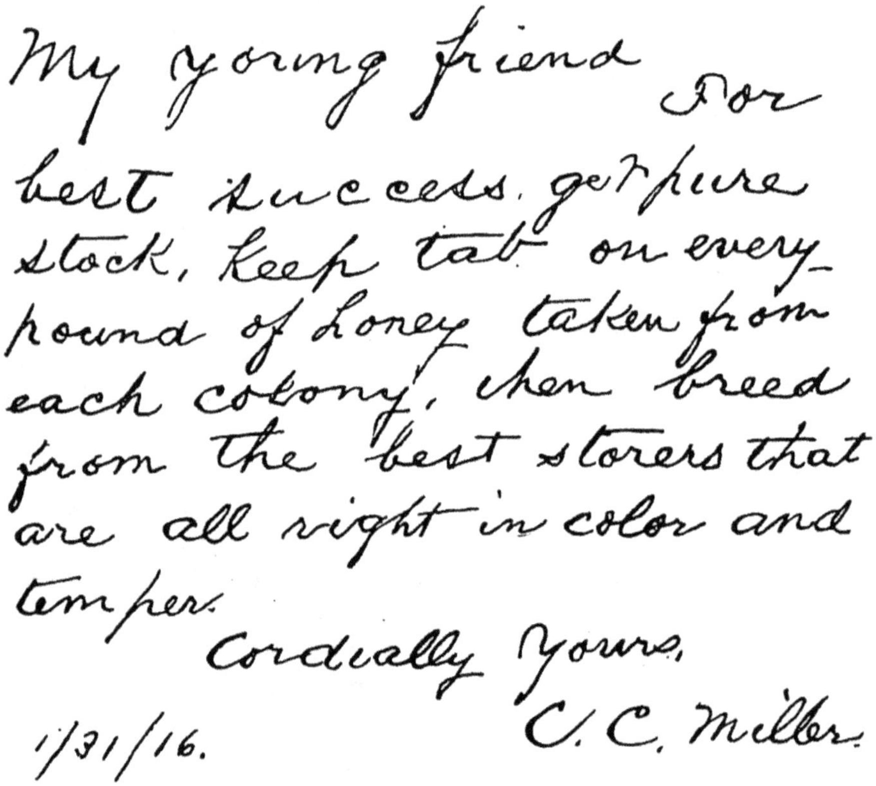

My young friend

For best success, get pure stock, keep tab on every pound of honey taken from each colony, then breed from the best storers that are all right in color and temper.

Cordially yours,

C. C. Miller.

1/31/16.

Fig. 3. The most important thing in beekeeping is the Queen.

sider the most important thing in all bee-culture, if you consider one of any more importance than the rest?

A. A thorough knowledge of everything connected with the business. Perhaps you want to know which is the most important, the bees, pasturage, hive, or some other thing. Hard to say.

Bees are no good without pasturage, and pasturage is no good without bees. You can't very well get along without a hive. But if you insist that I must pick out some one thing to which the beekeeper must give the greatest attention, I think I would say the queen. For whatever the queen is, that decides what the bees are. By breeding for the best all the time, a man is more likely to get ahead than by giving his attention to something else, such as hives or pasturage.

Bees, Cross.—Q. I have a colony of bees that is very cross, and one that is very tame. How could I introduce a queen from the tame colony to the cross one so as to make them all tame? And at what time ought I do it?

A. Rear a queen from the better stock, kill the objectionable queen, and introduce the new queen in an introducing cage. Or you may do the other way. Take two or three frames of brood from the good colony, put them in an empty hive, fill out with empty combs or frames filled with foundation, and set this on the stand of the bad colony, moving the bad colony to a new place close by. Now lift out two or three frames from the bad colony (be sure you don't get the queen), and shake the bees from these frames into your new hive, returning to the bad colony its two frames of brood. In two or three weeks there ought to be a queen laying in your new hive. You can strengthen it by adding brood and bees from the bad hive, or you can unite with it all of the bees and brood, killing the bad queen two or three days before uniting. Pehaps you would like to have two colonies instead of one. In that case kill the bad queen a week after the first move, and two or three days later exchange one of the two frames in your new hive for one of the frames in the bad hive, making sure there is a queen-cell on the frame, and also on the frame you leave.

Bee-Escapes.—Q. When you have on more than one super how would you put a bee-escape under? Would you lift the supers one at a time and put them on a bench, and then, after the escape is on, put them back?

A. If there are two or more supers on the hive you are not likely to want to take all off at a time unless at the close of the season. So lift off supers until all are off that are ready to take, setting them on end on the ground, leaning against their hive, or perhaps setting them on top of an adjoining hive. Then return any that are not ready yet, put on the escape, and then the super or supers that are ready to take.

Q. Does the Porter bee-escape ever get clogged up with bees trying to carry out dead bees, larvæ, etc?

A. Yes, although there is not much chance for it. Dead bees are not likely to be in supers, neither is brood often present.

Q. Will queens and drones pass easily through the Porter bee-escapes?

A. Not nearly so easily as the workers.

Bee-Houses.—Q. I propose to build a bee-house in the spring, for protection against too hot summers and the cold months of winter. Our summers are not long, but sometimes very hot; the winters short and not very cold, occasionally in winter the thermometer will fall as low as 15 degrees above zero. Kindly give your advice on this question, also the advantage or disadvantage.

A. Bee-houses, such as you contemplate, were more or less in use some years ago, but have been mostly abandoned. They have the advantage that when the bees are handled in summer they will not sting so much as out doors, and they are safer from thieves. But they are hot and inconvenient for the beekeeper. In spite of the fact one does not generally relish advice against one's own inventions, I advise you to let the bee-houses alone.

Bee Hunting.—Q. How can I find bees out in the woods?

A. Set your bait and watch the direction the bees go when they leave it. Then move your bait in that direction, and try again. Keep on till you find that the bees go back in the opposite direction, and then you'll know you've passed the right place, and you can bait back nearer to it; all the while keep close watch on the trees to see or hear the bees flying in or out. Another way is to cross-line. After watching the direction the bees take, instead of moving directly in that line, move at right angles to it and watch the line the bees make. Now guess about where the point would be where these two lines cross, and try accordingly.

Q. (a) What kind of bait is the best for lining bees in the woods?

(b) How can I set it so the bees will scent it?

A. (a) Honey diluted with water, perhaps half and half. Some make a smudge by burning, and some flavor the bait with anise. Some make a smudge by burning old combs.

(b) Set it out in the open in the woods where the bees are prospecting.

Q. (a) What is a good way to hive a large swarm of bees from a bee-tree? The small entrance is about 20 feet from the ground, and the tree is too valuable to be cut.

(b) When is the best time?

A. (a) I have some doubt whether there is any way by which you can get those bees into a hive—provided the tree is not to be cut—without costing more trouble and labor than the bees are worth. Possibly you might smoke 'em out, if you can in some way secure footing enough to operate so high up in the air. The first thing is to decide as nearly as you can where the colony is located with reference to the entrance, for I take it from what you say that there is only one entrance. That may be at the top of the cavity, at the bottom, or somewhere between. With your ear against the tree, listen to the noise of the bees when you pound upon the tree, and you may be able to locate them. If the entrance be at the top, or near the top, then make another hole at the bottom; otherwise make a hole at the top of the cavity. Then into the lower of the two holes send something whose odor will drive the bees out of the upper hole; carbolic acid, tobacco smoke, etc. Even ordinary wood smoke from a smoker may suffice if persisted in. As soon as the bees are out., plug the holes so they cannot return, and then treat them as a swarm.

(b) If you want to save the bees, a good time is not later than fruit bloom. If you want merely to get the honey, take it at the close of the honey-flow.

Bee Martins.—Q. Do martins seriously bother bees? If so, would they prove a handicap to a person who is just starting bee-keeping in a community where there are a great many of these birds?

A. I have never heard that martins were seriously troublesome to bees.

Beemoth.—Q. How does the beemoth get a start? It seems to start after combs are taken off the hive.

A. The beginning is an egg laid by the beemoth, and this hatches out into the larva, or "worm," as it is commonly called, in which state it does its mischief in destroying honey combs, after which it changes into the moth.

The trouble seems, as you think, to be worse **off** than **on** the hive, because off the hive there are no bees to protect the combs, although the eggs are generally laid on the combs while they are still in the care of the bees. It seems strange that the bees will allow moths to lay their eggs in the hive, but they do. At least black bees do, to some extent, although Italians seldom allow it.

Q. What can I do for worms in bees?

A. The best remedy for wax-worms, as the larvæ of the bee-moth are called, is a big lot of bees. The worms are not likely to get much of a start in a rousing colony, but a weak, discouraged colony is their proper prey. If your bees are blacks, you will find that changing to Italian blood will be a great help. Indeed, a colony of good Italian blood, even if quite weak, will keep the worms at bay.

If the worms have made a fair start, it may be worth while

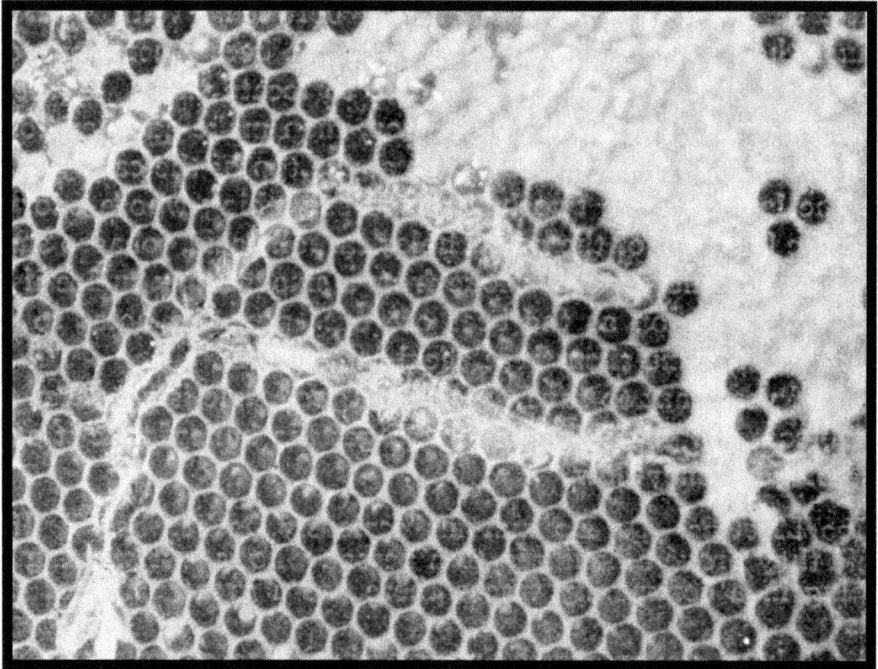

Fig. 4. Tunnels of the moth in a brood-comb.

to give the bees some help. At least you can dig out the big fellows. Take a wire nail and dig a hole into one end of the gallery that the worm has built. Now start at the other end, and as you dig the gallery open the worm will crawl along and come out of the hole you first made, when you can dispatch it.

Q. How early in spring will the wax-worm begin its destructive work on combs stored in the honey-house?

A. Something depends upon the character of the honey-house. It needs considerable warmth for the favorable development of the miscreants, and if your honey-house is a warm place you may expect them to flourish by the first of May. Otherwise not till the last of the month. In a cool cellar there will be little

trouble before the combs are needed for swarms. Of course, if
the weather is warm their work will be earlier than when there
is a cool spring. Moths cannot winter in a house where it
freezes hard.

Q. Do bees carry moths while swarming?

A. I don't believe that bees ever carry with them the moth,
its larvæ, or its eggs.

Q. (a) I had two weak colonies which I was going to unite,
but found a weavy web on the combs and in them a handful of
small worms. Those on the comb were about three to the inch in
length, and not a live bee to be found, and no honey. The worms
resembled cut worms.

(b) Is that comb of any use to put in other hives?

(c) How did the worms get in the hive without the bees destroying them?

A. (a) The worms were the larvæ of the beemoth.

(b) Yes, unless too much of it is destroyed.

(c) Eggs were laid in the hive by the moth, and from these
eggs worms were hatched. The colony must have been weak and
like enough queenless.

Q. I have a lot of honeycombs that I will have to keep
through the summer months. What is the best remedy to keep
the moths out of them? I have them packed closely in a chest.
Will fumigating them with sulphur do, or is bi-sulphide of carbon
the best?

A. Sulphur will do, but it takes a gread deal of it to finish the
big worms, and it does not kill the eggs, so that it must be used
again two weeks later to kill the worms that have hatched out
from the eggs that were left. Carbon disulfide (which is the
later name of bisulphide of carbon) acts more vigorously, and at
one operation cleans up big and little, eggs and all. After you
have the worms all killed you must keep the combs where the
moth cannot get at them.

On the whole, it is nicer to give such combs to the bees. They
will clean them up and keep them in nice condition. You can fill
a hive-body with them and put it under a colony, so that the bees
must pass through in going out or in.

Q. I have a number of frames which look very ragged on account of moth ravages, some in which more than half of the comb
is gone. Will the bees repair this and fill out the frames again
if I give them to the bees next spring? Or would I better cut
out all this comb and put in new foundation?

A. If the comb is in good condition except for the ravages

of the moth, it's good property, and·it is well worth your while to keep it to give the bees again. Something, however, depends upon how the bees fill out the vacancies in the combs. If they fill them up with drone-comb you might better melt up the combs

FIG. 5. What remains of a comb devastated by beemoth.

and give foundation. If given to a strong colony in a flourishing condition you can count on a lot of drone-comb; if given to a nucleus, or to a swarm when first hived, you may count on worker-comb.

Q. If I brush the bees from my section honey and put it in folding cartons, such as are listed in supply catalogs, right in the bee-yard, will I be bothered with the beemoth in my honey, and will this not save trouble in fumigating? Of course, this honey will be well sealed before putting in cartons.

A. No, you can't trust to anything of the kind. Years ago, if I took off sections and kept them where no moth could touch them, within two weeks tiny worms would appear here and there. The only way I could understand it was that the moth must have gotten inside the hive and laid the eggs on the sections. Of late years I have no trouble of the kind, probably because of the Italian blood. With black bees I had a good deal of trouble, and fumigated with sulphur. Carbon disulfide may be better.

Bee-Paralysis.—Q. Two of my colonies are killing what

seems to be old bees. They turn black, and are driven out. The brood seems to be all right. I had one colony affected the same way last year that became all right. I would like to know the cause and cure, if any. I have over 100 colonies, but never saw anything like it before.

A. The probability is that it is a case of bee-paralysis. The bees are black and shiny from losing their plumage. They come out of the hive and jump around on the ground, generally with bodies somewhat distended, and there is a peculiar trembling motion of the wings. The sound bees appear to pester and drive the sick ones. As far north as you are, it is doubtful if you need pay any attention to it. I've had several cases of the disease, and never did anything for it and the disease disappeared of itself. Far enough south it becomes a terror, and although many cures have been offered they generally fail to effect a cure. O. O. Poppleton says he cures by sprinkling sulphur over the bees and comb. Texas beekeepers of late claim that excessive dampness in the hive is the prime cause. They practice shaking the bees onto perfectly dry combs in a dry hive.

Q .Will camphor prevent bee-paralysis if I put a small piece in the hive?

A. It will probably have no effect whatever.

Bees, Best Strain.—Q. What is the best bee for this country, the Buckeye strain, 3-banded, golden Italian or leather colored?

A. There are good bees of almost all kinds; the majority of beekeepers probably prefer the 3-banded leather colored Italians.

Bees Dying.—Q. What ails my bees? Quite a number of them are dead or dying. One day when the snow was on the ground I saw dead bees on the snow. While I was there a bee came flying out of the hive, lit on the snow and was frozen; it was zero weather. I have a box set over the hive; the front side is open. They are not packed. The entrance of the hive is wide open, and they have plenty of honey to winter on, with nothing to disturb them. They are Italian bees.

A. There may be nothing wrong at all; depends upon what is meant by "quite a number." In a strong colony it is nothing strange if a thousand bees die off in the course of the winter; and when the sun is shining upon the white snow it is not alarming to see a bee fly out to meet its death in the snow.

Q. What is the cause of a colony of bees dying in the winter with plenty of honey in the hive? It seemed to be in good shape when it went into winter quarters.

A. It may be that the cluster of bees was in the center with honey on both sides; the honey was all eaten out of the center, and the bees drew to one side; they ate all the honey on that side and a long cold spell prevented their going to the other side until they starved to death, leaving plenty of honey in the hive; or, the colony may have been queenless and weak at the end of the honey season.

Bees Flying Out.—Q. On a warm afternoon the bees crowd around the entrance to the hive almost clogging it full. Some of the bees crawl up the front of the hive, take wing and fly back to the entrance of the hive and go in. Some of them crawl up the front of the hive and fly away. A few take wing from the entrance and after circling back and forth in front of the hive, fly away. Why is this?

A. The young bees are taking their play spell and at the same time marking their location. You will notice that at first they fly with their heads toward the hive.

Bees, Livelihood from.—Q. I have been trying to decide on a move for several years; that is, in the keeping of bees. I had a slight experience of two years with bees, but just became greatly interested in them when I left the country to accept a position in the Postal Department in New York City. I still hold such a position, but my desire and love for bees have increased so much that I am contemplating a change to the country. My hesitation comes from the doubt whether I could make a good living from them alone should I devote my entire time to them. What is your opinion? Would it be wise and profitable to give up my position of $100 a month to lurch into beekeeping? I would not go in extensively at the start, but try and feel my way as I advance. Will you kindly give me advice I seek as to whether there is a profitable field in the keeping of bees as a business proposition?

A. Your question is one that is exceedingly difficult to answer. If it be a mere matter of dollars and cents, I should say that beekeeping is a good business to let alone, for the same amount of brains and energy that will make you a living at beekeeping will make more than a living at almost any other business. But if you have the great love for beekeeping that some men have, then it may be the part of wisdom for you to choose beekeeping in preference to any other business that would net you ten times as much money. For your true beekeeper doesn't have to wait until he has made his pile before he begins to enjoy life, but every day is a vacation day, and a day of enjoyment.

But you must have a living. Can you make a living at beekeeping? I don't know. There are a few who make a living at beekeeping alone. There are probably a few more who can. You may be one of them, and you may not.

It would not be advisable for you to cut loose from everything else and start in at beekeeping with the idea of making a living at it from the very start. If you have enough ahead so that you can afford to do nothing for a year or two, with a fair assurance that you could take up your old line of work at the end of the year or two, if you should so elect, then all right. For you must count it among the possibilities that the next two years may be years of failure in the honey harvest.

If you can take such a risk, perhaps you can grow into quite a business with bees, while still continuing at your present business. Indeed, that might be the best way. In a suburban home you could probably care for 25 or 50 colonies mornings and evenings. Or, you might have a roof apiary in the city. The profit from them would be all the while bringing you nearer the point when you could cut loose from everything else. After a year or two you could judge better than anyone else whether it would be feasible and advisable to try beekeeping alone.

Bees Restless in Winter.—Q. I have two colonies of bees I moved '14 miles last December. I packed them in chaff about 3 inches thick, and they have plenty of honey. They seem restless and come out of the hive when it is 20 degrees below zero. What is the cause of this? Are they too warm?

A. The likelihood is that not very many bees are coming out, and a very few need cause no alarm. If the number is considerable it may be that a mouse in the hive is disturbing them, or that they are troubled with diarrhea. In the latter case a good flight the first warm day will cure them, unless, indeed, they have unwholesome stores, which will keep up the trouble more or less until warm weather comes.

Beespace Over Brood-Frames.—Q. I build my own hives. Is it necessary to have beespace between cover and brood-frames? I find some hives do not have this.

A. By all means have a space of about one-quarter inch between cover and top-bars. This for the sake of allowing a passage over the frames in winter, and also because if there is no such space the bees will glue the cover tight to the top-bars. This is on the supposition that there is nothing between the cover and top-bars. With some people it is a common practice to have a sheet or quilt over the top-bars, and in that case no space is needed except enough room for sheet or quilt.

Beestings.—Q. If you wash yourself with salt and water before handling bees, will it help to keep them from stinging?

A. Unless your hands are dirty, I don't believe washing in salt water will do any good, and then soap is better than salt. When bees are swarming they seldom feel like stinging.

Q. What is the best remedy for a beesting, either for a person on whom the sting swells or on one on whom it doesn't? It does not swell on me. I have heard that a sting will always swell on a healthy person. Is that true?

A. To give all the remedies that have been offered for beestings would occupy pages. Perhaps as good as any other remedy is a plaster of mud. Most beekeepers of experience seem to think that no remedy does much good; the only thing they do being to get the sting out as soon as possible. Don't pull the sting out by grasping it between the thumb and finger, for that helps to squeeze more poison into the wound; but scrape it out with the finger nail, or else, if it is in the hand, by striking the hand hard upon the thigh with a sort of sliding motion, which wipes out the sting. A sting will swell on a healthy person in nearly every case if the person is not used to it, and perhaps a little worse on an unhealthy person; but after being stung often one generally becomes to an extent immune, so there is little or no swelling. Among remedies offered are ammonia, salaratus of soda, juice of lemon or plantain leaves, kerosene, cloths wet in cold water, etc.

Q. Does the sting of the honeybee ever prove fatal? I have heard that if a person is stung on the end of the nose it is fatal. Is this a fact?

A. I don't believe a sting of itself ever caused a death. There have been cases where persons died after being stung. I've been stung many times on the nose, and I'm not at all dead.

Beeswax.—Q. What is beeswax, or what does it originate from?

A. Look closely at a lot of bees, especially at swarming-time, and you will see some of them that have, along the underside of their abdomens, little plates of pure beeswax, somewhat pear-shaped. That's where the wax originates, being secreted by the bee from the food it eats, somewhat as the cow secretes milk from the food she eats.

Q. Since (as it would seem) no established beekeeper produces enough wax to work into his necessary foundation, where does the surplus come from?

A. "Things are not what they seem;" at least not always. An

established beekeeper may not produce enough wax for his own
foundation, and again he may. If he works for extracted honey,
and has reached the point where he makes no more increase and
needs no more combs, he may have a surplus of wax from his
cappings, and probably will have; even if he renews his combs,
the melted combs should furnish wax for the new ones. Upon
him the comb-honey man may depend for his wax. There are
also beekeepers who use little or no foundation, and such men
are likely to produce surplus wax by means of the combs they
melt up from the diseased colonies.

Q. What is the best method of producing beeswax? I want
beeswax instead of honey. (New York.)

A. So far is I have ever learned, those who make a business
of producing wax rather than honey have done it by feeding back
the honey, thinned, as fast as the bees built combs and stored it.
But that was in places very far from market, where the honey
would not pay for transportation and wax would. It is not likely
that you can make it pay in your region.

Q. Please tell me how I can purify beeswax. I can melt it and
get it out of the combs by the hot water process, but after I get
it melted I cannot get the dirt separated from the wax, as underneath the wax there is some kind of fine dirt; that is, the dirt
does not settle to the bottom of the vessel that the water and
beeswax are in. I would like to know some way to get this dirt
out of the wax, and will you please give me a way to mould the
beeswax into one or two pound blocks?

A. Your wax is only following the general rule. A large part
of the impurities, while heavier than wax, are lighter than water,
so they settle between the water and the wax. In other words,
you will find a layer of sediment on the under surface of the cake
of wax when it cools. There is not much difference between the
weight of the wax and the sediment, so that it takes it a long
time to settle. So if the wax cools very rapidly, much of the sediment will be mixed up with it. Your effort must be to keep the
wax in the liquid state a long time; or, as it is often expressed,
you must let the wax cool slowly. One way to do this is to cover
up warm with blankets or something of the kind. If the amount
of wax is small it will be longer cooling if you have a good deal
of water under it. Another way, with a small amount, is to put it
in the oven of the cook-stove, leaving the oven door open until
the fire begins to die down in the evening, then shut the door and
leave it until morning. Put the stove handle in the oven, and

then in the morning you will not forget to take out the wax before building the fire.

Then you scrape the dirt from the bottom of the cake, which you can do more easily while the cake is a little warm. With a large amount of such scrapings it may be worth while to melt the whole to get out the little wax in it, but with a small amount it is not worth the trouble.

Q. Is beeswax injured by coming to a boil? If so, can it be detected that the wax has been boiled?

A. Bringing to a boil will hardly hurt it if not repeated too much, nor continued too long, and I don't believe the short boiling could be detected.

Q. Will a brass or copper vessel injure the quality of the wax?

A. I think not. Iron will darken it.

Q. How many pounds of honey does it take to make one pound of wax?

A. For a long time it was counted 20 pounds. Then some figured it out 7 pounds or less. Possibly 10 or 12 pounds may not be far out of the way.

Beeswax, Rendering (See Combs, Rendering.)

Bee Supplies, Ordering.—Q. I have 12 colonies of bees in good movable-frame hives. I am a beginner. What shall I order in the way of supplies? I wish to run for comb honey and increase by natural swarming. I have nothing in the way of tools, and my time is limited, as I am a rural mail carrier.

A. It is not an easy thing to tell what anyone needs without pretty full particulars as to harvest and conditions. In general terms I should say that you should have on hand enough sections all ready in supers in advance, so that you can give to the bees as many as they would fill in the best season you have ever known, and then an extra one for each colony besides. Possibly you have had so little experience that you don't know what the bees would do in the very best kind of a season. Well, then, we might guess that in the very best kind of a year you would get an average of 125 sections per colony, although that may be putting it pretty low if you are in a good location. If your supers hold 24 sections each, as a good many supers do, it would take about five supers to hold the 124 sections, as we don't need to be so exact about it. But some colonies will fill more than the five, and some less; you can't hold them to the exact number, and at the last there will necessarily be more or less unfinished sections on the hives

when the season closes; so you ought to count on an extra super
for each colony; altogether six supers per colony, or 72 supers of
sections for the 12 colonies. Understand only once in a while you
will have a season when you will need so many, but you never
know but what the next season will be a bouncer, and you must be
prepared for it. What are not needed will be all right for the
next year. Even if the season proves an entire failure your supers will be all right for the first good season that comes.

As to hives, you will probably want to double your number,
preventing all afterswarms, so you will need to have in readiness
a hive for each colony, or 12 in all.

Bee Veils.—Q. What is your idea of a good bee veil?

A. Our favored veil is made after this fashion: One end of
the veil is sewed to the outer brim of the hat (of course, an elastic
may be used to slip over the hat if preferred); this keeps the veil
smooth, avoiding wrinkles in front of the face. An elastic cord
is run in the lower hem. A safety pin is caught through the hem
in the front, taking in the elastic cord. This is always left hanging in the veil, then when hat and veil are on, all that is needed
is to pull the elastic down until taut—not only taut, but stretched
very tight—and then to fasten the safety pin to keep it so. If a
rigid cord were used instead of an elastic, when the body was
bent it would become slack and allow bees to pass under, but if
the elastic is drawn down tight enough no bee can get under, no
matter what change is made in the position of the body. Nothing
can be simpler as a fastening, and it is perfectly safe.

Beginning in Beekeeping.—Q. I would like to know the best
possible way to commence beekeeping the coming season.

A. Take my advice and don't wait for the coming season, but
begin now, getting a good book on beekeeping and studying it
thoroughly. That's the way to begin, and by the time you have
done that you will know plenty well the next step. Begin with
two or three colonies, so that you may learn as you go.

Q. I am in Northern Minnesota. Does it make any difference
whether I get my bees from the Southern States, or would it be
better to get them as near home as possible?

A. Better get them as near home as possible. Transportation
from any distance south would be more than the cost of the bees.
If you can't do any better, get black bees in box-hives, and then
you can transfer and Italianize.

Bisulphide of Carbon (See Carbon Disulfide.)

Black Bees.—Q. I bought some bees last year for pure Italians, but now there are black ones in the hive. Could they have been pure Italians? This is my first year with bees. (July.)

A. You do not say whether there were any black bees in the hive last year. If the workers were all properly marked last year, it is possible that the queen was superseded last fall or this spring, and that the new queen is mismated. If there are only a few black bees in the hive, they may be from other colonies; for bees do more in the way of shifting from one hive to another than is generally supposed. Look in the hive and see whether there are any black bees among the downy little fellows that have just hatched. If there are, then either the queen has been changed or the queen you bought was not pure.

Q. Does the black bee enter the supers more readily than the Italian?

A. I think so; but I have no trouble with Italians.

Blackberry.—Q. Does blackberry yield much honey?

A. Blackberry is not generally in sufficient number to count much. I don't know for certain, but I think it might be important where there are large fields of it. In any case, whatever it does yield is of importance because it comes early enough to fill in the gap between fruit tree bloom and clover.

Bottom-Boards.—Q. As I expect to make my bottom-boards. I would like to know how deep an entrance can be before the bees will build comb from the bottom-bars to the bottom-board. I have been using seven-eighth inch deep, but I notice the bees alight outside and crawl in, the same as a three-eighth inch entrance; but if one and one-quarter inch they don't alight on the bottom-board, but on the combs. It seems to me that this must save quite a little of the bees' time. Would two inches be too deep?

A. I don't know just how deep a space would do, but I'm sure 2 inches would be too deep. I have had bees build comb in a one and one-quarter inch space, although from what you say your bees may not yet have built in such a space. I should feel safe with a three-quarter inch space, and likely there would be little building with a space of one inch. My bottom-boards however, are all 2 inches deep; then during the busy season I fill half or more of the space with a sort of rack, which prevents the bees building down, yet gives them the chance for much ventilation. In winter the bees have the whole 2-inch space which is an important advantage.

Q. Which is better, to have the bottoms loose on the hives, or have them nailed on?

A. The best way is to have the bottom fastened to the hive by means of staples, so that you can remove it at any time you like. I wouldn't have a bottom that could not be fastened on, and a bottom that couldn't be taken off readily would be worse still.

Q. How would it work to use the same depth of bottom-board under the frames, seven-eighths in winter, and close the entrance down to three-eighths inch by a strip of wood for out-door wintering? What size of entrance would you use here?

A. It would be all right. Deeper than seven-eighths would be still better for the bottom-board, but I would not care to have the entrance more than three-eighths, and perhaps not more than four inches wide.

Bottom-Boards, Dust on.—Q. I have noticed on the alighting-boards of two or three of my colonies a substance resembling sawdust. What is this? I winter my bees outside in small sheds packed with straw. The sheds face the south.

A. That brings vividly to mind the first year I wintered bees, when I was alarmed to find under the bees and at the entrance something that looked like a mixture of coffee grounds and saw-dust, and I didn't know but what it was "all up" with my bees. An old beekeeper quieted my fears by telling me it was nothing worse than the bits of the cappings that the bees dropped when unsealing the honey. Your bees have the same "disease."

Box Hives.—Q. In June I found a large swarm of bees and put them in a shoe-box, not having any beehive. I have left them in the shoe-box, and I think there must be about 100 pounds of honey in it, as it is all that I can do to lift it. What is the best way to get a portion of this honey without damaging the bees or their winter supply? What is the best way to keep bees over winter? My cellar is rather cold, and slightly damp. Would it do to keep them there? (Illinois.)

A. It is very doubtful whether you can take any honey away without badly damaging the chances of the bees for safe winter-ing. Better leave it until spring, or until next summer, after the bees have swarmed. They will not waste it, and you can get later what honey they can spare. If they were in a movable-comb hive you could safely take the honey now.

You are in latitude 41 north, or a little more, and in Illinois that's nearly the dividing line between outdoor and cellar winter-ing, with mostly cellaring. But if your cellar is damp and cold, and there is no way to warm it, you may do better outdoors.

Brace-Comb.—Q. What is a burr-comb? What is a brace-comb?

A. The terms "burr" and "brace" are used somewhat indiscriminately, "burr" more properly referring to bits of comb built over the top-bar or elsewhere, perhaps without connecting two parts together; and "brace" being used to designate bits built between frames or combs, thus serving to brace them.

Brood Carried Out.—Q. I have 13 colonies and there are five of them that are carrying out brood almost ready to hatch. What is the trouble, and how can I prevent or remedy it?

A. Late in the season drone-brood will be thrown out at a time when bees kill off drones. A very few specimens of such brood thrown out at any time may have no special significance. There are other causes. In the spring of the year the bees are likely to use large quantities of stores in rearing brood, and it generally takes several years for a beginner to learn that unless they have a big lot of honey on hand there is danger of starvation. You have on hand, perhaps, a plain case of starvation, and of course there is just one way to prevent or cure, and that is to feed. Occasionally a very fertile queen will lay faster than her small colony can care for the brood. Cold, raw days may chill brood not properly covered. This brood will die and be carried out. These instances are rather rare, and the amount of chilled brood is usually small.

Brood-Chamber Clogged With Honey.—Q. "Why don't bees go into supers?" Brood-chambers are clogged with brood and honey, and "nothing doing" in the supers. Advice given is to uncap the honey in the brood-chamber. Most of the sections have bait-comb in them. I have no uncapper, so I have run a hook over the capped honey and considerably disturbed it. Now, how about being as sure as possible that in these hives with clogged-up brood-frames (with honey) there will be enough bees growing in September or August so as to have the colonies winter all right? Is there such trouble in producing extracted honey? What had I better do?

A. Running a hook over the sealed surface ought to have somewhat the same effect as uncapping, but is probably not as good. If you have no regular uncapping-knife, a common butcher-knife will do fairly well. When the surface has merely been scratched I have known the bees to repair the capping, not taking up any of the honey. But if the knife cuts down to the honey, they are bound to take up some of the honey before they can do anything at repairing the capping, and if everything is full below they are to a certain extent compelled to deposit the honey above.

If a good fall flow comes, that may start an increase of brood-rearing, and the bees may empty some of the honey from the brood-chamber into the super. If no fall flow comes, there is danger, as you suggest, that brood-rearing will be so limited that the colony will not be so strong for winter. Yet there is this crumb of comfort in the case, that if there is nothing for the bees to do in the field they will not grow old so rapidly, and will not die off so fast, so that, after all, they may not be so very weak for winter.

With extracting-combs there is less inclination to cram the brood-chamber, yet if the bait-sections be as fully drawn out as the extracting-combs the difference should be very little, unless it be that extracting-combs that have been used as brood-combs have greater attraction for the bees than comb that has never had anything but honey in it.

Are you sure that your colony has enough ventilation so that the bees may be comfortable in the upper story?

Brood-Chambers, Two-Story.—Q. Would you approve of the plan of using two brood-chambers, one on top of the other, to enlarge the brood-nest of 8-frame dovetailed hives? I do not wish to keep more than 6 or 8 colonies, but I would like to keep them strong.

A. Decidedly. It often happens that before the clover harvest is over, a good queen will be hampered in a single 8-frame hive, and then it's a good thing to add the second story. But if you are working for comb honey you should reduce to one story at the time of putting on supers. That can hardly be said to be reducing the room of the colony—merely giving the room in the super in place of the brood-chamber.

Q. At about what date do you contract from two-hive bodies to one, and do you do it by shaking? I assume that most of the workers must be left in the single chamber, so that the gathering force there may be as strong as possible.

A. You will find in "Fifty Years Among the Bees," that about as soon as clover-bloom begins, or at least within ten days after seeing the very first blossom, I give section-supers, and at that time I reduce to one story, leaving in that one story all the brood I can, and all the bees, shaking or brushing all the bees from the combs I remove.

Brood Dead.—Q. This fall we doubled up on a few colonies by putting the weaker colony on top and a sheet of newspaper between. When I took the frames out of this top hive to hang them in the basement for the winter, I found dead brood in them. I thought perhaps this brood was not properly taken care of by the bees. There was no smell, the brood was not ropy, and the unsealed brood was coffee colored, while the sealed was white and thin.

A. Like enough the few bees deserted the brood and went below, leaving the brood above to starve.

Q. I have one colony that is carrying out brood in all stages of development, some alive with wings almost developed. Can you tell me what is wrong?

A. One guess is that the larvæ of the bee-moth, or wax-worms, have mutilated the young bees with their galleries and the bees carry them out. Another is that the bees are driving out the drones and destroying the drone larvæ, or they are starving.

Brood-Frames.—Q. Should the honey of the brood-frames be extracted, and can it be done without injuring the brood?

A. Unless you are very careful you are likely to throw out brood if any is in the comb; and it is not considered best to extract honey from such combs.

Brood-Rearing.—Q. If there is plenty of honey, at about what time do bees quit rearing brood?

A. Somewhere about October 1, some earlier and some later, depending upon age of queen, condition of colony, and part of country. In the south, brood is reared practically the entire year.

Q. I notice that late brood-rearing is recommended. How would you encourage it?

A. With a fall flow of even moderate extent there is no need to do anything to keep up late brood-rearing. Young queens, however, are more reliable than old ones. If the flow stops early, breeding can be kept up by light feeding every other night.

Q. Why do bees rear brood in December and January? They have very little honey. (West Virginia.)

A. It is nothing unusual for bees wintered outdoors to begin rearing brood in February, especially as far south as Virginia, and not very unusual in January. I think December is unusual, and I don't know why any of yours should begin so early. Possibly there is something in their condition causing it.

Q. Will a colony rear brood in February or March if it has been given frames of sealed honey in the fall? I gave it outside frames, which, I don't think, had any pollen in them, or beebread.

This swarm I caught late in August, so it did not have time to procure stores for the winter.

A. A good colony wintered outdoors will be likely to rear brood before February is over, if it has pollen. If no pollen is present, you need not expect brood till pollen can be gathered.

Q. Will bees rear brood sooner in spring when wintered in the cellar or on the summer stands?

A. They begin rearing brood, as a rule, sooner outdoors than in cellar. Even in the north, brood-rearing outdoors begins often, if not generally, in February, and in the cellar generally not till March.

Q. What is the most satisfactory way of stimulating brood-rearing in the spring?

A. The most satisfactory way for me is to see that the bees have plenty, yes, more than plenty—abundance—of stores; keep them well closed up, and then let them entirely alone. If your queens are not so good at laying as to do their best without the lash, or if your locality is such that you have good flying weather without any pasturage, then it may pay you to feed half a pound of diluted honey every other evening, or to change end for end the outside comb on each side.

Brood Scattered.—Q. When you find little patches of brood deposited here and there in the combs, what does it indicate?

A. Probably a failing queen.

Brood, Spreading.—Q. What are the indications when it is safe and profitable to spread the brood, i. e., place an empty comb in the center?

A. For some years I have been of the opinion that for me there is no time when it is profitable to spread brood. Early in the season, at the time when we want bees to build up as fast as possible, the bees of their own accord have all the brood they can cover. In that case, if brood is spread it can result only in chilled brood, thus hindering instead of helping the building up. I don't know whether the bees of others are different or not. If at any time your bees are covering combs that have no brood or eggs at the outer part of the cluster, it ought then to be safe and profitable to spread. But be sure you're right before you go ahead.

Brood Uneven.—Q. I have one colony of bees whose cells are uneven on top—some tall and some low. What is this? Some of the brood looks pink, but does not smell. I have a virgin queen in the hive. Could she be a drone-layer only, or not purely mated?

A. That is not the work of a virgin or unfertilized queen, but rather of an old queen. It is nothing very unusual when a queen becomes quite old for the store of spermatozoa to become to a certain extent exhausted, and then some of the eggs laid in worker-cells will not be fertilized and will produce drones, and

FIG. 6. Drone and worker-brood, irregular; showing the work of an old or inferior queen.

the cappings of these will be raised. It is not the work of laying workers, for in that case none of the brood would be sealed level.

Buckeye.—Q. Is buckeye honey bad for bees?

A. I never heard it was.

Buckwheat.—Q. (a) Does buckwheat bloom at the same time that white clover does?

(b) How much should be sown to the acre?

(c) Does it make the bees want to swarm in the fall?

(d) Is the grain good for chickens?

(e) Is buckwheat honey better than clover?

A. (a) No; buckwheat is much later, usually being sown after clover is in bloom, say about the last of June.

(b) Some sow two pecks to the acre some twice as much.

(c) It is not likely to make the bees swarm.

(d) The grain is good for chickens.

(e) No; it is dark, strong, and generally sells for considerably less than clover, yet some prefer it.

Q. Can I sow buckwheat in the spring, and continue at stated times through the summer, so as to have it bloom at certain periods, and make it profitable?

A. It is not advisable. Buckwheat seems to fit better as a late growth. Even if it should succeed when grown early, it would not be desirable where the earlier harvest gives honey of a lighter color, and better quality.

Q. I have a patch of buckwheat now in full bloom, but my bees do not pay any attention to it. What do you think is the cause of this?

A. I think buckwheat sowed about the first of July yields nectar better than that sowed earlier, and yours may have been sown too early. However, buckwheat is like white clover and other plants, it sometimes fails to yield nectar, no matter whether early or late, and I don't know why. Bees rarely work on buckwheat bloom in the afternoon.

Buildings, Bees in.—Q. There are a few swarms of bees in a house, and one in the bank building, which are troublesome. The openings are very small. How can they be killed?

A. Try putting in four or five tablespoonfuls of carbon disulfide (called also bisulphide of carbon.) It must be done at a time when all the bees are in, some sort of a crooked funnel being arranged to make the liquid enter the hole, and the hole promptly closed. Have no light near, for fear of an explosion.

Q. I have an old frame building and between the walls honeybees have made a home. There are three or four colonies in this building, and I would like to know if it would be possible to get them out from between the walls and put them in standard hives?

A. Cut away the walls so that you can get at the combs, and put them in the hive; leave the hive as near as possible to the old place of entrance; close up the wall so no bee can get into it, keeping the bees smoked out until this is done; then gradually move the hive each day to where you want it. That's the general principle, which may be varied according to circumstances.

Buildings for Bees.—Q. Is it possible to keep several colonies of bees in a building, using a window as a common entrance for all?

A. Yes, with proper precautions. The room must be light enough so bees can easily find their own hives after they are in the room, or else a tube for each hive to the outside, and there must be no chance for a bee to get out of the tube into the room. In the first case (the light room) precaution must be taken against bees flying against the glass where they cannot get out. The window, or windows, must have an opening at bottom and top of each window, or no sash at all.

Bulk-Comb Honey.—Q. Please explain the Texas method of having comb honey in jars. What is your idea of it?

A. I suppose you refer to the bulk-comb honey produced in Texas. Get honey filled in frames, cut out the comb, pack it in jars or cans, and fill up the interstices with extracted honey.

If I were in Texas I'd gladly go in for bulk-comb honey.

FIG. 7. Bulk-comb Honey. The comb is cut and placed in the can or jar, the interstices being filled with extracted honey.

Q. Are starters or full sheets of foundation put into frames every time the full combs are cut out when running for chunk honey?

A. Some use full sheets, some use starters, and some use neither, when the comb is cut out leaving enough of the comb under the top-bar to serve as a starter.

Q. How would it work for bulk-comb honey to put on an extra body of Hoffman brood-frames with brood foundation, or would it be better to use section foundation?

A. The thinner foundation would be better for table honey, and yet some have reported that the heavier foundation was thinned down by the bees. It would not be a bad plan to try each, and then you would know better what to do in future.

Bumblebees.—Q. Do bees and bumblebees ever sting each other to death?

A. I think it is not very uncommon for a bumblebee to attempt to enter a hive and to be seized by the bees. I have seen such cases, and oftener I have seen the dead body of a bumblebee at or near the hive entrance, the hairs stripped from its body. I have an impression that the honeybees are never stung by the bumblebees, although the honeybees often sting the bumblebees.

Burr-Combs.—Q. Should burr-combs be cut out from between frames when they appear?
Would bees tear them down as they do queen-cells?

A. It is better to cut them out every year or two, as they are in the way, and make it difficult to crowd the frames together without killing bees.

No, the bees never clean out burr-combs, and the presence of any of them between frames seems to be an invitation to the bees to build more. On the whole, it may pay to clean them out every spring.

Q. How can I prevent burr-combs?

A. You cannot prevent burr-combs entirely, but you will get along with a minimum if you will avoid too large spaces wherever burr-combs are likely to be built—don't have spaces more than one-quarter inch.

Buying Bees.—Q. When would you advise to buy bees?

A. Rather late in the spring, say about the beginning of fruit-bloom, is a very good time. The troubles resulting from wintering are likely to be over then, with nothing to hinder a prosperous career.

Q. I want to start an apiary and don't know where to obtain some Italian bees. Will you please give me the desired information?

A. I have no means of knowing any better than you. Your first effort should be to get the bees as near by as possible, since expressage is very expensive, and the railroads will not accept bees by freight. A little ad in your local paper might discover someone close by having Italian bees, of whom you had no knowledge. Possibly you may find in the advertisements of the bee journals what you want, and if not, then an ad in a bee paper, costing very little, would probably bring a number of offers.

Lately pound packages are largely used, though hardly advisable for beginners.

Cage for Introducing.—Q. What cages are the best for introducing queens? What kind of candy is used in them?

A. Merely for introducing without shipping, the Miller cage, with Scholz or Good candy.

Campanula—Campanilla.—Q. What is the Campanula, where does it grow, and is the honey from it of good grade?

A. Campanula, or belleflower, has not any reputation as a honey plant. But the Campanilla blanca of Cuba (ipomæ sidæfolia) also called "Aguinaldo de Pascua," is one of the principal honey plants of Cuba. There are several varieties. The honey is light, about like white clover, and is said to have a very fine flavor.

FIG. 8. The Campanilla in full bloom.

Candy for Bees.—Q. As I have some colonies of bees light in stores, how can I make candy out of granulated sugar to carry them through the winter?

A. You can make Scholz or Good candy, but the probability is you have not the extracted honey, so all you need to do is to make just plain sugar candy. Into a vessel of boiling water on the stove, stir two or three times as much sugar, and let it cook until a bit of it dropped into cold water appears brittle; then pour it out into greased dishes so as to make cakes half an inch to an inch in thickness. These cakes may be laid on top of the frames and then covered up any way to keep snug and close, so the bees will go up to them; for if too cold the bees will not leave the cluster to reach them, and starve with abundance in the hive. Then promise yourself you'll not be caught that way again, but will have plenty of combs of sealed honey each fall to meet any emergency.

Q. How can I make queen-candy for introducing cages?

A. Heat a little extracted honey (don't burn it), and stir into it some powdered sugar. Keep adding all the sugar you can until you have a stiff dough. Even after you seem to have it quite stiff

FIG. 9. "Campanilla Blanca" and a sealed frame of its honey.
"The famous honey of Cuba"

you can still knead in more sugar. Then let it stand a day or so, and very likely you can knead in a little more sugar. No danger of getting it too thick. You will notice that no definite quantities are given, but you will use several times as much sugar as honey. At a rough guess I should say that if you begin with one spoonful of honey you will have five spoonfuls of candy. Of course, if at any time you should get in too much sugar, you can add honey. It is not really necessary to heat the honey, only it hurries up the work a little. Government regulations require that honey used in candy for mailing cages be first boiled in a covered vessel, to kill germs of bee diseases.

Q. Why is it that hard sugar candy is used as winter feed while the candying of honey in the hive is deplored? Why not feed candied honey over the cluster when needed?

A. Your question is hardly a fair one, for it sounds like saying that there is no objection to feeding candy, while there is objection to letting the bees have candied honey. The fact is that there are good authorities who deplore the feeding of sugar candy more than the candying of honey. There is, however, not so much said against the feeding of sugar candy, because it is often a choice between that and starvation, in which case the feeding of candy is not a thing to be deplored. In the case of honey candying, it is to be deplored, because it is not so good as liquid honey. It remains, however, to say that it is quite possible that it is better to feed candied honey than to feed sugar candy, and that so good authorities as the Dadants have practiced feeding candied honey. Perhaps ye Editor will tell us about it in a bracket.

(Sugar may be **crystalized** in lumps like rock candy, in which case it is of no use to the bees. But soft candy makes good bee food. The same may be said of granulated honey. If the honey has granulated in a way that there are hard, crusty lumps in it, some of it may be lost by the bees, especially if they attempt to consume it in dry weather. When the atmosphere is loaded with moisture, much of this softens so the bees can use it. But well ripened honey which has a soft granulation will be consumed to the last mite. We have often fed candied honey in the way suggested by our correspondent.)

Candying (See Granulation.)

Cappings.—Q. I have been told that yellow flowers tend to make cappings yellow, too, or, in brief, that the bees will cap honey from yellow flowers with a yellow capping. If this is true, please explain.

A. Yes, it is true, at least of some flowers, dandelion, for example. I suppose the bees get the yellow coloring from the pollen.

Q. What methods, if any, besides the knife, have been used since the invention of the extractor to get rid of the cappings of the combs?

A. Turn to page 306 of the American Bee Journal for October, 1908, and you will find description and illustrations of the Bayless uncapping machine. Several other machines for uncapping have been invented, but none absolutely perfected.

Q. To melt up cappings and wax scraps, what would be the simplest way to do?

A. Use a solar wax-extractor, or hold your refuse till you have a sufficient quantity to send to a dealer who makes a business of wax rendering. You may also render the cappings in an ordinary wash boiler, with water.

Q. (a) How can I dispose of water which is a little sweet so as not to have the bees bother?

(b) Is there any chemical or other article which can be mixed with the washings of wax and cappings to be thrown out that will not attract the bees?

A. (a) I have never paid any attention to it, for if it is thrown into a drain or upon the ground it is so diluted that it disappears before the bees pay any attention to it. If you find the bees trouble in that way, you could add more water to it before throwing it away, so as to make the sweetness very slight, and then if each time you throw it on a new place on the ground, I think you will have no trouble. The reason for extreme caution in the matter is the fear that there might be foulbrood in the honey. You might make a sure thing of it by having a pit dug, into which you would throw the washings, and have the pit covered well. But why not save your sweet water for vinegar making?

(b) Carbolic acid would, no doubt, be effective.

Q. When the cappings of brood-cells are sunk, is it always an indication of disease, or are the cappings of healthy brood sometimes sunk?

A. I don't know that the cappings of healthy brood are ever sunk, but dead brood might have sunken cappings without any disease being present; and, of course, dead brood could hardly be called healthy brood.

Cappings Melter.—Q. Is there any melter that will do fast work and not injure the flavor of the honey that goes through it?

A. Maybe; but as the Scotchman says, "I hae ma doots." To do fast work there must be considerable heat, so that at least a little of the fine flavor would be hurt.

Carbon Disulfide.—Q. What quantity of carbon disulfide should be used for a stack of eight 8-frame supers of combs, and how often should the application be made to ensure against wax-worms?

A. Four tablespoonfuls ought to suffice. One application is sufficient, unless fresh eggs are laid in them again by the beemoth.

Carload of Bees.—Q. How many colonies can be shipped in a car? Please give instructions as to how to prepare bees for shipping, and how to load them in a car. Can the hives be placed on top of each other in the car?

A. Perhaps 500. The hives must, of course, be fastened bee-tight, but with plenty of ventilation. The hives must be placed so the frames will run parallel with the track, so the bumping will strike the combs lengthwise and not sidewise. If there are no more than will stand on the bottom of the car—say 90 to 100 hives —the only fastening needed will be strips nailed to the floor, so the hives cannot move in any direction the strips being one or two inches thick. If the hives are piled on top of one another, then they must be strongly fastened by braces running from side to side, or else from top to bottom, perhaps both.

Carniolans.—Q. Describe the color of the Carniolan bee. Some say that this kind of bee has yellow bands, and others say there are none of yellow color, but that they are all gray.

A. I think you would not recognize any yellow in Carniolans. They have very distinct rings on the abdomen, but these rings are not yellow, but whitish. They look much more like blacks than Italians, but have the credit of being a little larger than blacks.

Q. Is the quality of honey of the Carniolans better than of other races of bees? I have been told so, and that they don't gather honeydew as much as the other bees. Is there any truth in the statement?

A. The quality of honey gathered from the same source will no doubt be the same, no matter what bees gather it. I don't know whether Carniolans are less inclined than others to gather honeydew. I didn't know it was claimed for them.

Q. I have some Carniolan bees in 8-frame hives. If I had them in a larger hive would they swarm less? Can I get surplus honey in bigger hives; that is, if I have bigger brood-chamber? I like the Carniolan bees; they stand the severest winter and breed up faster in the spring. They gave me a nice surplus of honey early in the season, two supers to each colony. I sold the honey and got a good price for it. I have some colonies that will give four supers, and this is not the best honey year for Illinois, either.

A. Yes; a large hive will reduce the probability of swarming, since a crowded condition of the brood-nest is one of the chief factors in producing the swarming fever. Neither will a larger hive take away your chances for getting surplus. Formerly I used 10-frame hives, and changed to 8-frame hives chiefly because it was the fashion. If I were to start in afresh I would study some time before I would decide to adopt the smaller hive. With the larger hive I got fine crops of beautiful sections, and you can do the same.

Carniolan bees swarm very readily.

Catclaw.—Q. From what section is catclaw honey obtained? Is its honey of good flavor?

A. The catclaw, though found in most southern states, is a honey producer especially in Texas. The blossoms come out in late spring, the tree being low, bushy and spiny. The honey is only fair in flavor as compared to the whiter honeys like alfalfa.

Catnip.—Q. Is catnip honey fit for table use? It seems to taste very strong. (Iowa.)

A. Catnip honey has the reputation of being fine for table use. Unless you have catnip in great abundance, and little or nothing else yielding at the same time, you can hardly be sure that you have pure catnip honey; and it is possible that the very strong taste comes from some other honey being mixed with the catnip.

Q. Will catnip sown now, or in the spring, afford bee-pasture next season?

A. Catnip is a perennial, continuing permanently when once started. I think you cannot count on bloom the first season, but it will increase in size and strength after the second year. It seems to have a partiality for hedge-rows, but that may be because of the protection, for it grows well in the full blaze of the sun.

Caucasians.—Q. What color is the Caucasian bee, if the stock is pure?

A. About the same color as the common black bee.

Q. What are the physical features that distinguish the Caucasian bees from the Carniolans?

A. The main difference in appearance is that the whitish ring is not so distinct in Caucasians as in Carniolans. Carniolans look enough like blacks to make it hard to distinguish them, and Caucasians look still more like blacks. The rings on the abdomen of Carniolans are whitish; on Caucasians grayish.

Q. Are the Caucasian bees as gentle and as good honey-gatherers as the Italians?

A. The Caucasians were heralded as the most gentle of bees. No doubt some of them are, but others are reported as being no gentler than Italians. Not as much has been said about their gathering ability as their gentleness, and it is hard to say just yet what their status will be in that particular. They are still more or less on trial.

Q. I have 20 colonies of bees that I want to breed up to Italians or Caucasians. Which would you advise me to breed them up to?

A. Opinions differ; but Italians are so generally preferred that you will be safe in adopting them.

Cedar.—Q. My bees gathered pollen today (March 4), from red cedar ,and as I have failed to find cedars referred to as a source of honey or pollen, it struck me as something rather unusual. Is that a common occurrence?

A. It is quite likely that it is nothing unusual, even though no mention may have been made of it. It is only the plants from which unusual quantities of either honey or pollen are obtained that are generally mentioned as honey-plants.

Cellars for Bees.—Q. I wish to build a bee-cellar to hold 200 colonies. I intend to build it in a side hill and have it entirely under-ground, and cover it with a roof, then 3 feet of earth, then a roof over this to keep the earth dry. How large would you build it?

A. Ten cubic feet for each colony is not far out of the way.

Q. Would you make any special arrangements for ventilation? If so, how would you arrange the ventilators? The sides and ends will be built of stone and mortar.

A. It might be a good plan for you to have a ventilator, because it is easy to provide one when building, and not so easy afterward, and if you find you are better off without it you need not use it. T. F. Bingham had a repository not so entirely under-ground, and he believed in a ventilator 16 inches square. A plain board pipe from near the ground up will answer.

Q. My cellar floor is concrete and is always damp. I am thinking of covering it with four inches of dry sawdust. What do you think about it? I wintered 106 colonies in this cellar without a loss in 1914, but the covers and bottoms were very damp in the spring. I gave the bees all the ventilation possible, according to the weather, and the temperature stood at about 45 degrees on top and 42 degrees below.

A. I'm afraid the sawdust will not do a great deal of good. Possibly it might if you should sweep it up and dry it out as fast as it got wet. Lime might do more good.

Q. My cellar (28x30 feet) has a hot water boiler in it. The temperature varies from 48 to 52 degrees. Do you think I can winter a dozen colonies of bees in it successfully?

A. It ought to be a capital place. Without letting light into the cellar you should keep it open enough to have the air always fresh, not cooling it below 45 degrees.

Q. Will a cellar containing vegetables, potatoes, etc., be a good place to winter bees? Would the odor from the vegetables hurt the bees?

A. That depends. If, like too many cellars, with a lot of decayed vegetables and the air foul and moldy, the bees will not do

FIG. 10. A bee-cellar conveniently located to the apiary. The upper part is used as a honey and super storage room.

well. If the cellar is kept as clean as it should be for civilized beings to live over, the bees will not object to the odor of the vegetables.

Q. Is there any way to keep a cellar dry enough for bees when the thermometer is only 38 degrees, Fahr.? I lost all my bees the last three winters. I think it must be because of the dampness and the cold. What can I do to keep it warm and dry? We have had bees for the last twenty years, and have been successful until the spring of 1908, when we lost all.

A. Putting lime in the cellar will help to keep it dry. But at 38 degrees, the cold may be more to blame than the dampness. For years, before there was a furnace in my cellar, I kept a small stove in it, and kept a low fire in it whenever necessary to keep the temperature up to 45 degrees. It seems a little strange that after 17 years of success you should have a failure three years in succession. Like enough the tide will now turn, and you will again have good success. In my earlier years of beekeeping, I had experience as bad as yours, but by sticking to it I've made quite a lot of money from the bees since.

Q. If bees are put into a cellar under the kitchen, would the noises incident to the kitchen-work—running a washer, bringing in wood, constant walking, etc.—be a detriment to the bees, provided the hives were not jarred by any of these various operations?

A. I cannot speak with entire positiveness; but I have never noted any bad results from noises overhead (although I never had anything very bad in that line), and never heard of it from others, so I don't believe you need take into account the matter of noise, but put your bees in the place that gives you the best temperature and ventilation, providing there is any difference.

Q. Suppose a cellar is full of bees. Is it good or bad for the health of those who live in the rooms above the bees?

A. That depends on the beekeeper. If he's a poor beekeeper he will likely have a cellar with foul air and dead bees, and his cellar will be bad to live over. If the beekeeper is all right, the cellar will be kept clean, with pure air. The air in my cellar is as good as, or better than, the air in the living-rooms, for the cellar door is more or less open nearly all the time.

Cellaring Bees.—Q. Will you give me some light on how to carry bees into the cellar without the bees flying out and stinging? For years it has been a mystery to me how to carry bees in, and sometimes out, without closing the entrances. Is there a difference in bees, handling, location, or what? I am curious to know.

A. I will tell you just as nearly as I can just how my bees were carried into the cellar last year. They were carried in, November 25, in the morning. The cellar had been wide open the night before. Although that does not make much difference at carrying in as it does at carrying out, still it is better to have the cellar cool, so the bees will settle down quietly when brought in. The average distance of the hives from the cellar door was about ten and one-half rods. Then they were carried a rod or so further to their place in the inner room. Two able-bodied men took about two hours to carry in the 93 colonies. One of them was experienced at the business; I think the other had never carried bees before. Each man picked up his hive, carried it in his arms into the cellar and set it in its place. You may judge of the quietness of the bees when I tell you that no sort of protection was used in the way of gloves, veils or smoke, and the entrances were left wide open. There was one exception; I had failed to staple on the bottom-board of one hive, and when the bottom dropped off I had to use smoke to fasten it on. But I must hasten to add that last year was exceptional. I think they were never carried before without veil or gloves, for at least a few colonies would prove troublesome. I don't know what made the difference. Perhaps the bees were in an unusually dormant condition.

I am unable to say why your bees should act so differently. Some bees are more irritable than others; but I doubt if your bees are worse than mine in that respect. Perhaps one secret is in having the bees undisturbed for a long time before they are carried, and then being set down so quickly that they do not have time to get fully waked up. When they are in the most quiet condition it takes two or three minutes to get them thoroughly aroused, and in that time they are in place in the cellar. If they are stirred up ever so little, they are easily stirred up a few minutes later.

Cellar Wintering (See Wintering.)

Cells, Kinds of.—Q. Describe and tell how to know a queen or drone-cell from a worker-cell.

A. Lay a rule on the cells. If they measure five to the inch, they are worker-cells; if four to the inch, they are drone-cells; if larger and shaped like an acorn cup, they are queen-cells.

Cell-Protectors.—Q. How are those little cone-shaped wire queen-cell protectors used? I have a number of them, but did not use them because I could not make them cover the cell as I thought they ought.

A. The protector must be large enough to cover the whole of the cell after all superfluous wax has been trimmed from the cell. The point of the cell is put in the hole of the protector, and the four points of the wire-cloth twisted together. A slender wire has one end fastened to the protector and the other end of the wire is fastened to the middle of a nail. Two frames are pulled apart, the cell is let down between them, so that the cell will be at the middle of the frames, and the nail across the top-bars prevents the cell from falling down. Generally, however, there is no need of the nail, for the frames come near enough together to hold the cell; the wire serving to hold the cell in place till the frames are shoved together.

Cement for Hives.—Q. Do you think it would be all right to make supers out of cement? Would it agree with the bees? I can make them much cheaper than with lumber. (Arizona.)

A. My guess would be that cement supers would be quite objectionable on account of their weight. It is also possible that with 115 degrees in the shade they might be too hot.

Q. How about concrete for hive-bottoms? I am setting some of my bees on blocks I make for them right on the cool ground. Can they be used for winter?

A. They will probably work all right for either summer or winter. Of course, it would seem as if concrete would give the bees "cold feet" in winter; but then, they don't need to put their feet on the concrete.

Chaff Hives.—Q. I wish to work up to about 20 or 25 colonies of bees and have no bee-cellar to winter them in. I think of using nothing but chaff hives. Would you advise me to depend entirely on such hives?

A. I hardly dare advise. Chaff hives will make you less trouble preparing for winter, but they are cumbersome and unwieldy, and if they should perchance at any time pass into the possession of someone having a cellar or wanting to take them to an out-apiary, they would be objectionable. So it would not be a bad plan for you to experiment a little trying both kinds, only be sure to have only one size of frames. In northern latitudes the chaff hive is counted valuable.

Chestnut.—Q. How would chestnut lumber do for beehives?
A. From my recollection of it in boyhood, I should call it bad.

Chickens Eating Bees.—Q. Is it a common thing for chickens to eat bees? We had one that would stand in front of a beehive and eat bees until it was full. I thought it would die, but it kept it up for two weeks, and was doing well.

A. Testimony is somewhat mixed on this subject. For the most part it is claimed that chickens do not eat bees, or if they do, it is only the **drones**. Some, however, say that chickens eat workers, especially some chickens that have learned the trick.

Chunk Honey (See Bulk-Comb Honey.)

Cleome.—Q. What about artificial pasturage for bees? Is

Cleome pungens worth cultivating for the honey alone? Is the honey of good quality? Is it light or dark, and how does it compare with white clover honey?

A. Cleome pungens is not worth cultivating for honey alone. I do not remember to have seen any statement as to the character of its honey, and I don't know whether anyone ever secured enough of it to tell just what it was like.

Clover, Alsike.—Q. There is not a half acre of white or alsike clover within three miles. Would it be profitable to sow 15 or 20 acres?

A. Just for the honey alone that the bees would get, it would not pay at all. But if you take into account the additional gain from hay and pasturage, it might pay well.

Q. Does red or alsike clover bear pollen, or is it an excess of nectar that blights the seed when the bees do not gather it?

A. Red and alsike clover yield both nectar and pollen, but honey-bees do not often work on red clover. An excess of nectar would do no harm; but if the clovers are not visited by insects, especially bees, there will be no fertilization, and so no seed. Red clover is mainly dependent on bumblebees for fertilization.

Q. Will alsike clover make bee pasturage in this state (Kentucky)? A very small amount has been sown in this country the last year or so, white clover being the principal source of honey.

A. I think alsike may be counted on as a good honey-plant wherever white or red clover does well.

Q. Do you think it will pay to buy alsike clover seed for farmers to sow within one mile—say 40 acres? Would it make any perceptible difference in the yield of honey?

A. Yes, or sell it to them at a discount.

Q. (a) When should alsike clover be sowed?
(b) How much seed should be sowed per acre?
(c) Should it be sowed by itself, or with any other crop?

A. (a) At the same time farmers in your locality sow red clover.

(b) About four pounds.

(c) Either way, a favorite way being to sow with oats. On rich ground, where the alsike would be likely to lodge badly in wet weather, a sprinkling of timothy is good.

Clover, Crimson.—Q. Is the spring the proper time to sow crimson clover? If so, what time in the spring?

A. If sown in the spring it should be as early as frost is well out of the ground; but oftener it is sown about the last plowing of corn, and not expected to bloom till the next year.

Q. Which is the better for bees, crimson or alsike clover?

A. All things considered, perhaps alsike.

Q. Does crimson clover bloom the first season after sowing?

A. Yes, if sowed early enough it may bloom the same year; usually not till the following year. That's in the region of 42 degrees north latitude; far enough south it might more readily bloom the same year after early sowing.

Clover, Red.—Q. What do you think of the long-tongued or red clover queens? I have seen them advertised so much.

A. I think there is an advantage, and there may be a very great advantage in long tongues. In actual practice, however, I have come to doubt whether it is still worth while to pay any attention to the length of the tongues. Breed from the stock that gives best results. Very likely that may in most cases give long tongues, but whether tongues are long or short, we want bees that will get the most honey. Unfortunately, the quality of tongue length does not seem always to be handed down to succeeding generations.

Clover, Sweet.—Q. When is the proper time to sow white sweet clover seed?

A. When farmers in your vicinity sow red clover, alsike or alfalfa.

Q. Which is the better honey producer, white or yellow sweet clover?

A. Probably not much difference in yield, but the yellow is reputed to be about two weeks earlier than the white. That makes the yellow more valuable in some places, and the white in others. Where white clover abounds, the two weeks earlier would be of no advantage, as it would come at the time of white clover, and if the yellow also closes two weeks earlier than the white, the white sweet clover would be of more value. In localities where there is lack of forage during the first two weeks of the yellow sweet clover, the yellow clover has the advantage.

Q. In reading American papers, I observe frequent references to sweet clover as a plant for bee-pasturage. Is it the same as white clover (trifolium repens), which is the staple bee-pasturage here during the summer months? (New Zealand.)

A. Oh, no; it's an entirely different thing, growing sometimes to the height of 8 or 9 feet, although 3 or 4 feet is a more common growth. The most common sweet clover is melilotus alba. It is a biennial, coming from the seed one year, blossoming the next, and then dying, root and branch. Even if bees have all they can do on white clover, sweet clover is valuable, because while it begins to bloom later than white clover, it continues much later, even till frost.

There is a yellow sweet clover (melilotus officinalis) which blooms two weeks earlier than the white. Sweet clover will grow where scarcely anything else will, as in a clay bank. It seems to flourish best, or at least to start from the seed best, on hard ground trodden by farm stock.

Q. Last year white sweet clover was everywhere; this year there is scarcely any. Why did it not grow again this year, instead of the yellow? The bees worked on the white all the time, and seemed to be crazy over it, but they paid no attention to the yellow.

A. Sweet clover is a biennial, growing the first year without blooming. Then after blooming and producing seed the second year it dies root and branch. So, if you sow seed one year and leave it to itself thereafter, the tendency would be to have bloom every other year. One yellow sweet clover (melilotus indica) blooms the first year. It is an annual.

Q. How far north and south will sweet clover thrive and do well?

A. I suppose if sweet clover may be considered as having any native place it is Bokhara, in Asia, about 40 degrees north of the equator. At any rate, it is called "Bokhara clover," and years ago that was the chief name for it. According to that, one would suppose that it would be at its best on the parallel of 40, which runs centrally through Ohio, Indiana, Illinois, Utah and Nevada. But it does not seem to be very limited as to its habitat. I think it succeeds about as far north as bees are generally kept.

Q. Is sweet clover tender, or hardy? Will it freeze as easily as does corn?

A. Hardy—very hardy. Sweet clover would only laugh at a freeze that would kill corn. I think I've known it to be killed only in two ways. One year I prepared a piece of ground in fine shape, sowed sweet clover with oats, and it made a fine stand. Next spring there wasn't a spear left. The ground was so nice and soft that it heaved and pulled up all the sweet clover by the roots. In the solid ground of the road-side I never knew it to winter-kill. Another year I had a piece mowed close to the ground when it had started from the seed and was nearly a foot high, and that finished it.

Clover, White.—Q. What kind of clover is the best for bees? (Iowa.)

A. In Iowa, probably, all things considered, no clover is more valuable than the common white clover. Very likely you have that without any sowing. If you want to sow any besides, try sweet clover, both the white and yellow variety. It blooms later than white clover.

Q. What time of the year does white clover bloom in this state? (Illinois.)

A. In the northern tier of counties it opens its first blossoms in the last of May or first of June, and earlier as you go south.

Clustering Out (See Hanging Out.)

Cockroaches.—Q. What can I do for roaches? They bother the bees by getting in the brood-chambers, on sections, and all over the inside of the hive.

A. I didn't suppose cockroaches would do any particular harm in a hive where there are bees. You can poison them with some of the special poisons sold for that purpose, or with any other poison, only you musn't poison the bees. Put the poison between little boards only one-eighth of an inch apart, or in some vessel with a one-eighth inch entrance.

Colony.—Q. How many bees are estimated to be in a medium populous colony?

A. At a rough estimate, perhaps 30,000.

Q. What would you call a good colony?

A. A colony that, in early spring, has brood in five frames, Langstroth size (17⅝x9⅛), each frame being three-quarters or more filled with brood, would be a fairly good colony; with six or seven frames of brood it would be a very good colony.

Color of Bees.—Q. What causes such a great diversity in color among the individual bees and also among the colonies in general whose queens are a mother and her daughters?

A. If you have a pure Italian queen, her worker progeny all having the same markings, and from her rear a young queen, and this young queen mates with a pure Italian drone, you may expect to find the same markings in the worker progeny of the young queen as are found in the worker progeny of her mother. But if this young queen mates with a black drone, then you will find the worker progeny different, some of it looking like black workers and some like Italian, and perhaps intermediate markings.

Color Sense of Bees.—Q. What is there about the color of clothes to make the bees quiet when handling them?

A. I don't know why it is; I only know the fact, that cross bees are not so likely to sting one with light as with dark clothing. I have worn a good many different pairs of mason's or painter's white overalls for the sake of avoiding stings. I don't think white clothing particularly appropriate to my style of beauty, and in going through town to the out-apiary, I'm not fond of appearing on the streets arrayed in white, but I'd rather do that than to take the increased number of stings with dark clothing. But, mind you, I get all the stings I care for, even with white clothing. If bees are cross, they'll sting through the whitest clothing.

Comb.—Q. From what do bees make comb?

A. From what they eat—honey and pollen— much as a cow makes milk from what she eats.

Q. Is the comb in comb honey injurious to a person's health? Most people when eating comb honey swallow the comb.

A. Beeswax is utterly indigestible. It is sometimes used to make corks for bottles containing acids so powerful that they burn up ordinary cork, and of course the weak acids of the stomach can have no effect upon it. I have seen something about its being melted in the stomach; but the heat in the stomach is many degrees too low to melt beeswax. Even if melted, it would still be as indigestible as ever. But lots of indigestible things are taken into the stomach that do no harm; and may do good. When comb honey is chewed with other food and taken into the stomach some claim that the finely divided portions of wax are a benefit. Certainly they are not likely to do any harm.

Comb Foundation.—Q. How many pounds of comb foundation would it take to fill one brood-chamber of 10 frames with full sheets? Also, to fill one super with 28 sections?

A. For ten Langstroth frames it takes about 1¼ pounds of medium brood foundation, and one pound of light brood. For twenty-eight sections it will take about one-quarter pound of thin super foundation.

Q. What kind of foundation is best to use in the extracting frames?

A. If you use shallow extracting-frames, you can use light

brood foundation, only you must be careful about turning the extractor too fast while the combs are new. Indeed you can use light brood with full-depth frames.

Q. Do you consider light brood foundation sufficiently heavy to be used with your splints in regular Langstroth frames?

A. Yes, only in place of five splints, as with medium, seven splints must be used with the light brood foundation. At least I did not feel safe to do with less than seven, and had good results.

Q. Is there any special width of foundation to use in a brood-chamber?

A. It is well to have the foundation come down to within one-half inch of the bottom-bar.

Q. I am going to buy foundation for 1,500 Hoffman frames for the next season and do not know whether to buy medium or light brood. I have used both and can see but little if any difference in results. I have had no trouble with light brood sagging. I wire my frames, but do not use splints. Which do you think is the better to use?

A. It's merely a question of which will succeed better, and as you have no trouble with the lighter foundation sagging, it will be economy to use that.

Q. What kind of foundation would you recommend for honey in one-pound sections?

A. Some prefer "extra thin," but, all things considered, my own preference is "thin."

Q. Which is the better to use in the frames, full sheets or starters?

A. Full sheets. If you use only starters you will have entirely too much drone-comb.

Q. Have bottom-starters of foundation in brood-frames ever been used?

A. I am not sure whether anyone else has ever tried them, but I have. But I had no use for anything of the kind after I found I could use full sheets of foundation clear down to the bottom-bar by the aid of foundation splints.

Q. I have about 40 pounds of light foundation, two years old, and it seems dry. Should I use it, or have it worked over?

A. It is all right to use as it is.

Q. In one place you say "Have the sections filled with worker-foundation," and in another place I read: "Dr. Miller has for years described his method of using bottom starters (as well as top ones) in sections of comb honey." Kindly explain this method, as I have never seen it in the papers. I have the Daisy foundation-fastener, and would like to try bottom and top starters. Would you make them meet in the center? Or how much space between the starters? When they are fastened only at the top, they twist and do not hang true.

A. The matter is very simple, and your Daisy fastener is just

FIG. 11.—A frame containing mostly drone-brood, the result of a narrow starter of foundation.

the thing to fasten a bottom as well as a top starter. It wouldn't do at all to let the two starters meet in the center, for in that case the bottom-starter would be certain to fall down and make a mess. When you buy foundation for sections, you are likely to get it in sheets 15½x3⅞ inches. This is just right to make four starters of each kind. The top-starters are 3¼ inches deep, and the bottom ones ⅝. For a section that is four inches deep inside, you will see that would leave a space of ⅛ inch between the starters. In reality the space will generally be more than that, for the hot plate melts a little of the edges of the starters. First fasten the bottom-starter, turn the section over immediately, and put in the other starter. If your bees are like mine, the first thing they will do on being given the sections will be to fasten the upper and lower starters together.

Even for the home market, I should prefer the bottom-starter. It makes a nicer looking section. - Unless a single starter comes down so far that it is likely to sag, some of the sections, especially when honey is coming in slowly, will not be built down to the bottom. Although the bottom-starter is original with me, I don't believe I'm sufficiently prejudiced in its favor to stand the extra trouble unless there were a sufficient gain to pay for it.

Q. How many sections 4¼x4¼ will one pound of thin super foundation fill; full sheets?

A. About 100, or a few more.

Q. Don't you think it would be a good plan for the manufacturers of foundation to furnish the **section** foundation with drone-size base? It would save the bees considerable work in comb-building where full sheets are used.

A. You would probably not like it. Generally there is less drone-comb in the brood-chamber than the bees would have if left to their own devices, and with little or no drone-comb below and abundance above, the queen would be likely to make trouble.

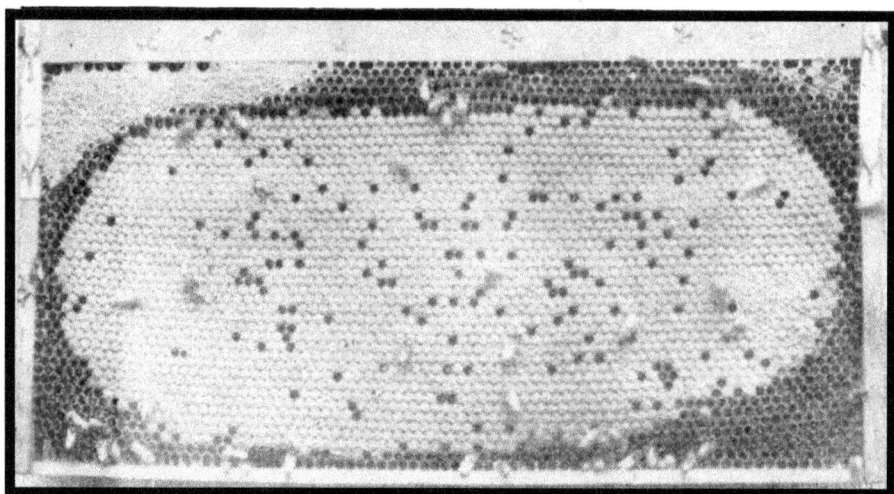

FIG. 12.—Full sheets of foundation assure combs with a minimum of drone-brood.

To be sure, you might keep her down with an excluder, but that would be trouble and expense, and you would find that some sections would not be finished up as promptly as they should be, for the bees would hold the cells open for the queen. I think, however, that if you care to try it you can get drone-foundation.

Q. If I order more foundation than I use, how can I keep it from spoiling?

A. I hardly know what you can do with it that it will not keep, unless you put it in an oven where it will melt, or spread it out in the sun and rain for a year. Just keep it covered up wherever it is convenient. Even if you have it filled into sections, keep them where they will be dry and nice, and they will be all right. Although bees take hold of fresh foundation a **little** more readily than that which has been kept over, there isn't much difference. But if you leave it on the hives in the fall, when no honey is coming in, it may become so bad that bees will not touch it next year.

Q. Please state the advantage in using the reinforced comb foundation. Some claim it takes less than other comb foundation, being thick on top and thin on bottom. If there is any comb foundation that is better please let me know, and if it is a fake, then also give the facts.

A. I did not know that it had ever been claimed that less foundation was needed if reinforced. Likely what you mean is that a less weight of wax might be used in filling a hive with foundation. I do not see why that may not be true. Foundation for brood-combs must be of a certain weight to prevent sagging. But the sagging is chiefly at the top. Now, if we use lighter foundation and reinforce the top part, there is a saving of wax. It is claimed, also, that bees begin work more promptly on the wax that is painted on. I have never used it enough to speak with authority, but I do not believe there is any fake about it, and I do not remember having seen a report from anyone who condemns it after having tried it.

Q. Is it not a fact that many combs affected with foulbrood and other diseases are rendered into wax, and that the foundation on sale by all dealers is contaminated more or less with this same wax?

A. Undoubtedly much wax is made from foulbrood combs, and just as undoubtedly much of it must fall into the hands of the manufacturers of comb foundation. But it does not follow that the foundation is contaminated so as to make it in the least dangerous. The continued high temperature to which the wax is subjected, when being made into foundation, destroys the spores. I think that some hold, too, that even if a spore were not destroyed by the heat, it would not germinate after receiving an impervious coating of hot wax.

Comb Foundation Fastening.—Q. Can full sheets of foundation be used for brood-frames without using either wire or wood-splints? Would it sag so as to spoil the cells for brood-rearing?

A. Unless the foundation be extra heavy it may sag enough to stretch a good many of the cells in the upper part.

Q. Please give me the method of fixing foundation (full sheets) in frames with wires; also starters, say 5 or 6 inches deep.

A. I may say briefly that if you have top-bars with kerf and wedge, it will be easy to insert the upper edge of the sheet in the kerf, and then push in the wedge deep. Then one of the ways of fastening the wires in the foundation is with the spur wheel, doing the work in a very warm room, so the foundation will be soft. If you have no kerf in top-bar, then run melted wax along the joint between the foundation and the top-bar.

I don't want to tell how to put in 5-inch starters, because I don't want you to use them. No economy in it. You will have entirely too much drone-comb. "You're going to use them anyhow?" Oh, all right, then. Put them in exactly the same as full sheets.

Q. Should comb foundation come close to the end-bar of the frame and be fastened there with wax? I wire my frames.

A. Either will do, but it is well to have the foundation come close to the end-bar. It is not necessary to wax it there.

Q. How far from the bottom-bar ought foundation to be?

A. About one-quarter of an inch.

Q. Can a sheet of finely woven wire be rolled between two sheets of wax in making the foundation for brood-combs, to take the place of splints or wiring frames, as now practiced? The sheets could be made the size desired, or the wire screen could be woven 1-inch mesh and could readily be cut out with the scissors.

A. Yes; such a thing has been advertised and in use for years, the Van Deusen flat-bottom, wired foundation, upon which there was a patent. Cloth and other materials, such as screen wire, have also been tried as a base for foundation, but in no instance were they successful.

Q. How do you fasten foundation sheets to the top-bars of shallow frames with no grooves and wedges?

A. With melted wax. Some use two parts wax to one of rosin. Make a board large enough to fit loosely inside the frame, nail stops on the ends so as to let the frame go down half way, put frame over, then the foundation in place, and pour the melted wax from a spoon with its point bent together, or else with a special dropper. The wax is likely to stick unpleasantly to the board unless you wet the board or else put newspaper over it. A brush may also be used to put on the wax.

Q. Is it necessary to wire full sheets in shallow extracting frames?

A. You can get along without wiring if you are careful.

Comb Foundation Drawn Out.—Q. Where foundation in sections has been partly drawn by the bees last year, will it do to use those sections and foundation this year, or had I better cut it out and put in new foundation?

A. If it is clean, with no remains of candied honey, use it again.

Q. How and when is it best to have brood-combs drawn out, or made from full sheets of comb foundation?

A. Give such frames of foundation any time when bees are gathering more than enough honey for their daily needs, if you think they will not stop gathering before they have time to finish the combs. Of course, that's as much as to say that the very best time is at the beginning of a harvest that you have good reason to expect will last two weeks or more. A strong colony, of course, will need less time than a weak one.

Q. Can I put frames with full sheets of foundation between two combs and get good worker-combs that are not stretched too much at the top? I mean without wiring.

A. You may, by using foundation splints or very heavy foundation. Even then you will not always get the best results between two drawn-out combs, for too often these combs will be bulged into the comb between them.

Q. Will bees draw out foundation as soon when it has been in the frame three months as they would if only in the frame three weeks? I like to put my foundation in the frames in the winter time, when I have plenty of time. This is to be the new foundation just made.

A. Speaking very strictly, I suppose the fresher the foundation is, the better. But I have used foundation that had been fastened in four or five years, and I've some question whether the bees made any great difference between that and that which had been put in only four or five days. At any rate, I believe it good policy to get it ready in advance, as you propose.

Q. A successful honey producer says full sheets of foundation are drawn down to the bottom-bar very much better when placed in a super than in the brood-nest. Is this so?

A. Sure.

Comb Foundation Gnawed by Bees.—Q. What is the reason that the bees gnaw the foundation starters in the brood-chamber? I have found two or three starters lying at the bottom of the frames. A few days later I found a strip that they had carried out in front of the hive.

A. The starter may have been insufficiently fastened; there may have been something objectionable about the foundation; it may have been that the bees were not gathering, and at such a time they will gnaw foundation as if in pure mischief. They may gnaw it to use the wax elsewhere about the combs.

Comb Foundation Splints (See Splints.)

Comb Foundation Making.—Q. I keep a few colonies of bees for my pleasure and have saved some wax. Now, I don't like to sell wax for 20 cents a pound and buy foundation for 65 or 75 cents. Can you recommend the Rietsche press? If not, say "no" to my second question; but if you can, please give a few hints as to how to make foundation. Are you making your own foundation? Could I make foundation? I have never seen it done.

A. My time has always been so fully occupied with other things that I never tried making comb foundation. Besides, I think I can buy it cheaper than I can make it. I use foundation mostly for sections, and it would take a great deal of practice to enable me to make anything like as nice foundation as those do who make a business of it.

There are thousands of Rietsche presses in use in Europe, and in the foreign bee-papers one sees nothing but praise for them. With the instructions that you would receive with the press you could probably succeed, even without ever having seen foundation made.

Q. What is used as a lubricant on the rollers of a foundation mill? The one I have sticks. I cannot set it close enough to make any cell-walls at that. This is the first time I am using it, as I had a lot of foundation bought shortly before buying the mill, so did not try it before.. I am using just clear water now. I dipped the sheets last winter.

A. Starch is used as a lubricant, also honey or soap. If your sheet of wax is too cold, the wax will not be pressed up into a side-wall. Try having the wax warmer.

Comb Honey, or Extracted.—Q. I have a few hives of bees and wish to increase, but am undecided as to which to do, buy fixtures for section, or extracted honey, and, if section, whether plain or beeway. It may save me quite an expense later on.

A. Whether it is better to produce comb or extracted honey depends upon the honey and the market. The darker honeys do not sell so well in sections, and in some places consumers prefer sections so strongly that even dark honey pays better in sections. From what I know of your location, I think you have light honey, but your market for extracted honey is unusually good, so that my guess would be that you would do well to extract.

Comb Honey.—Q. How much honey can one expect from a colony during a good season, provided no increase is made?

A. The amount varies greatly; from nothing to 300 pounds or more. Dr. E. F. Phillips estimates the average at 25 to 30 sections per colony. That, of course, takes good seasons with bad. If you take good seasons alone, it might be twice as much.

Q. What is the best way to handle bees in regard to room between the white honey-flow in June and July, and the buckwheat flow in August? My bees are then too strong to occupy an ordinary hive-body, and if given new sections they destroy the foundation and spōil the sections.

A. I don't like to get into a quarrel with you, but I am hardly ready to accept your statement that your bees are too strong to occupy an ordinary hive-body, and at the same time destroy foundation sections. Not but what that is true, too, but I don't agree with your evident belief that the bees need more super-room. If they tear down the foundation in sections, they are not gathering anything more than they need for their daily use, and so need no super-room.

"Can't stay in the brood-chamber?" "Let 'em stay out, then. Won't hurt 'em a bit to cluster outside the hive till it is time to put on sections for the buckwheat; this on the supposition that you want the buckwheat stored in sections. Another way is to give them a second story. If you haven't any extra combs to put in second stories, one or two combs in each story will be enough, so long as they are storing nothing, and you need not be troubled with the thought of the empty space in the upper story.

Q. How can I bleach comb-honey? I got about 2,400 sections last year, and it was hard to sell it on account of its darkness. I see a process for bleaching it in "A, B, C of Bee Culture," but do you know of any better way? All the honey that is coming into the market is whiter than mine, and I cannot account for it. If you know of a way to whiten honey, please let me know.

A. No; I can give no better way. It's one of the cases where prevention is better than cure, and I try to manage so there shall be as few darkened sections as-possible. There are two reasons for sections being darkened outside—being too long on the hive, and being too near old, dark combs. If a super of sections be left on the hive until every section is completely sealed, the central sections are very likely to be darkened. So don't wait for the sealing of all the sections, but take off the super when all but a few of the outside ones are sealed. Perhaps the four corner sections will not be finished, perhaps four on each side. Then these unfinished sections are massed together and given back to the bees to be finished. At one time, when I used wide frames to hold sections, my practice was to raise a brood-comb from the brood-chamber and put it between two frames of sections in the upper story, so as to induce the bees to begin work promptly. It was very successful in getting the bees to darken the capping of the sections, for they would carry bits of dark old brood-comb across to use on the sections, making them dark before ever the capping was finished. You will probably find that a thin top-bar will help to darken sections, because it allows them to be nearer the brood-combs. On that account a top-bar seven-eighths of an inch thick is desirable. You may also find more trouble with shallow brood-combs than the deeper ones.

The above refers to the color of the cappings. The honey itself may have been dark, perhaps honey-dew. There is no known process to change its color. As to bleaching the surface, some have reported success by simply exposing it to the light. A south exposure, allowing direct rays of the sun to shine upon the sections will work more rapidly than a north exposure, but care must be taken with a southern exposure, for in a place too confined, and with sections too near the glass, the heat might be so great as to melt the comb.

Comb Honey, Producing.—Q. Give the best method of working for comb-honey where the principal, and you might say all the honey-flow, comes between May 1 and 15. (Arkansas.)

A. The only special thing in such a case is to do your best to have all colonies strong early enough for the harvest. You will find that early in the season some colonies will be much stronger than others, and that the weaker colonies will be very slow about building up. Suppose you have some colonies with eight frames of brood, some with seven, some with six, some with five, some with four and others weaker still. You can take brood from any colony that has more than five frames, enough to reduce it to five frames of brood. Now, don't bestow that brood indiscriminately to the weaker colonies, but let the weakest wait till the last. Give a frame to each colony that has only four, and when these are all supplied, then help those that have only three, and so on. If all cannot be brought up in time, let it be the weakest ones that are neglected.

Comb Honey, Removing.—Q. Do you leave your comb honey all on the hive until the honey season is over, or do you take it off as fast as finished?

A. I take off a super as soon as it is all sealed except the corner sections, although often these will be finished, too.

Comb Honey, Shipping.—Q. Please give instructions how to crate and ship comb honey.

A. When you get the shipping-cases that are now furnished by supply dealers you will hardly need instructions for using them, for you can hardly case the sections wrong, they being so placed that one row comes directly against the glass so as to show the face of the honey. It is of first importance that this row next the glass be a fair sample of the whole case for the man who veneers by putting next the glass the best and inferior honey back of it, will in the long run be the loser by it.

Unless there be so large a quantity of honey that it can be fastened solidly in the car, it should be put in the crates sold by some supply dealers, the crates so placed that the ends of the sections shall be towards the front and rear, so as to stand the bumping of the cars. On the contrary, if the sections are hauled on a wagon, they should pe placed crosswise.

Q. When is the best time to ship comb honey?

A. Generally about as soon as it is ready. In very cold weather combs are in danger of breaking.

Comb Honey, Watery Cappings.—Q. What is the cause and remedy of comb honey having a water-soaked appearance? The cappings lie right on the honey. The honey tastes the same as any other, but it does not look as good as where the capping is pure white. I have a colony that produced 100 pounds more this season than any of the others, but a good many of the sections had this watery look.

A. You have answered the question yourself, when you say, "The cappings lie right on the honey." In other words, the bees fill the honey right up to the cappings, leaving no air-space between the capping and the honey. The remedy is to change the queen, or else use the colony for extracted honey. Any section may also acquire the same appearance after it is taken from the hive, no matter how white the bees made it, if it is put in a damp place. Honey is deliquescent, attracting moisture from damp air, and should be kept in a warm, dry place. Where salt will keep dry is likely to be a good place to keep honey.

Combs.—Q. How can I know the different kinds of combs?

A. The greater part of the combs in a hive you will find to be worker-comb, made up of cells that measure five to the inch.

Drone-comb is made up of cells that measure four to the inch. Generally you will find where the change is made from one kind to the other there will be a few irregular cells, called transition cells, Then there is also the queen-cell, still larger than either of the other kinds, measuring three to the inch. More nearly correct it is to say that a queen-cell is a third of an inch in diameter, for you never find a piece of comb made up entirely of queen-cells. Generally each queen-cell is by itself; and even if you find several queen-cells apparently close in a group, you will not find three such cells in the compass of an inch.

Combs Breaking.—Q. My bees are doing nicely now, but I have trouble with combs of honey breaking and dropping down, caused by the heat. I have covers on all the hives, but the sun strikes the hive front. Is there any remedy for this?

A. The probability is that two things were responsible for the trouble. One was that the entrance of the hive was too small, giving the bees too little chance for ventilation. The other was that there was too little chance for circulation of air about the hive; buildings, trees, or bushes preventing a free movement of air. Years ago I had combs melt down in a hive—I think I never had them melt down in any other case—and the sun never shone on the front of the hive, nor any other part of the hive. The hive stood in a very dense shade, a thicket of bushes on one side, and tall corn on the other. The entrance was not very large, but I think the combs would not have melted if the hive had stood out in the sun all day long, provided there had been full chance for the breeze.

Combs, Preserving.—Q. (a) Does it injure empty extracting combs to keep them where the temperature goes below freezing? (b) If not, would it be safe to stack then up in the yard with a sheet of heavy tarred paper between each super?

A. (a) The combs may be slightly cracked with very hard freezing, but that is a small matter compared with the advantage that freezing kills all the beemoth, their larvæ, and even their eggs. I should certainly prefer to have the combs exposed to freezing all winter.

(b) That will be all right.

Combs, Moldy.—Q. If empty drawn combs remain in the hives all summer, and the hives are clean, is there danger of the combs becoming moldy? If such hives were not used, would you close up the entrances to keep out moths?

A. No danger of mold unless you keep the combs in a cellar

or damp room. I've some question whether you can close the hives tight enough to keep out moths. They squeeze through a very small crack. But if the combs are in a close building the moths are not likely to find them. Yet it is a pretty safe guess that, if colonies died on them, the worms are there already. In that case, whatever combs cannot be put in the care of the bees should be treated with sulphur fumes, or, still better with bisulfide of carbon.

The moldy combs will be cleaned up by the bees when given them.

Combs, Old.—Q. Will combs that have had brood reared in them from one to three years spoil the color and flavor of honey if used for extracting-frames?

A. There may be a slight difference, but you probably could not tell the honey from that stored in newly built combs.

Q. When having old combs in frames taken from colonies that died during the winter, to what extent is it good practice to dig the dead bees out of the comb?

A. Brush off all the bees you can, hold the frame flat and shake vigorously, shaking some of the bees out of the cells; leave those that will not shake out for the bees to dig out; they can do it cheaper than you.

Q. I have some brood-combs; they are black. I also have some that the moths have been in, that I lost earlier. Are those combs any good, or had I better throw them away? I thought I could use them for natural or artificial swarms.

A. If not too badly torn by worms they are all right to use again.

Q. How many years of constant use for brood can worker-comb have without diminishing the size of the bees? I have read that the cocoons left behind imperceptibly diminish the size of the cells of the future occupants, and prevent the bees from attaining their full development and size.

A. I have combs that are 30 years old or more, and I cannot see that the bees reared in them are any smaller than those reared in new combs. I remember that one of the patient foreign investigators—a German, I believe, whose name does not now occur to me—took the trouble to measure the contents of cells in combs very old and new, by actually filling them with liquid, and he found that the old cells contained just as much liquid as the new. The idea that the cells become smaller with age has been taught faithfully for many years, and there are still some who advise that combs be renewed every four or five years, but I think the idea is based only upon theory. Without any careful examination one might easily conclude that as something more than was there before is left in the cell, every time a young bee is reared in it, the cell must necessarily become smaller. But examine carefully and you'll find that the diameter of the cell at its mouth remains the same. You will probably find that the bees gnaw out some of the cocoon at the sides, leaving it at the bottom. That, of course, will make the cell shallower, but to make up for that, the bees add fresh wax to the cell-wall at the mouth of the cell. If they add to the cell-wall at the mouth, that ought to increase the thickness of the comb, oughtn't it? Well, that's exactly what it does. Measure the thickness of a piece of worker-comb from which the first batch of brood has just emerged, and you will find it measures seven-eighths of an inch. Take one old enough, and it will be fully an inch thick, and you will find the septum one-eighth of an inch thick. The only practical danger is that if the combs get to be old enough the spacing from center to center may become too small; in other words, the space between two combs becomes smaller. Don't worry about good, straight combs being hurt with age.

Combs, Rendering into Wax.—Q. I have a lot of combs from hives in which the bees winter-killed; also from late swarms last year that starved out during the long, cold winter. How can I convert these combs into beeswax?

A. If you have enough to make it worth while, the best way to get the wax out of your combs is to get one of the wax-presses or extractors that will leave in the remains a very small amount of wax. For a very few combs, however, it may not pay to spend much, and the solar extractor will do. You may also get out a large per cent with a dripping-pan. Take an old dripping-pan (of course, a new one would answer), split it open at one corner, put it in the oven of a cook-stove, with the split end projecting out of the oven so that a vessel set under it will catch the dripping wax. Put a pebble or something else under the inside corner, so as to make the wax flow outward. If the comb be previously soaked with water several days, and a single comb at a time be laid in the pan, the wax will not be tempted to hide in the cups made by the cocoons. But it will be slow work. You may also break the combs up into bits, provided you can have them cold enough to be brittle, put them in a gunny sack in a boiler or other vessel on the stove, weight down the sack, working it occasionally with a stick, and skim off the wax as it rises. With old combs, in which many generations of bees have been reared, it hardly pays to render the wax without water, for a

FIG. 13.—A modern wax-rendering equipment.

great deal of it is soaked in the cocoons and cast skins of the larvæ. Soaking these in water first, prevents the wax adhering to the residue, or slumgum, as they call it.

But in these latter days it's hardly worth while for you to fuss extracting wax from your combs when you can send the combs to those who advertise to receive them and extract the wax for you, (and they'll get out more wax than you will), and allow you pay for the combs in wax or foundation.

Combs, Straight.—Q.   I have one colony with combs built crosswise. How would you manage to get them straight?

A. It may be part of the frames are straight and the others only a little crooked. In that case you might be able to cut away the attachments and straighten the comb into its own frame. If all the combs are very crooked, you may consider it as a box-hive.

Q. How do you get straight combs built? Last year I used full sheets of foundation. The frames were wired with four horizontal wires. Almost every one "buckled" between the wires, and they are a bad lot of combs.

A. I wonder if you didn't depend entirely on the wires. The foundation should be fastened securely to the top-bar, either by means of the kerf and wedge, or, what some think better in a very dry climate, waxing the foundation to the top-bar; that is, running melted wax along the edge of the foundation on the top-bar. But you will probably have less sagging of foundation if you use foundation splints that are described fully in this book, as well as in the book, "Fifty Years Among the Bees."

Combs, Weight of.—Q. How much will ten frames of empty combs weigh, new and old, size 17⅝x9⅛, top-bar one inch? How much wax will ten combs produce, if rendered?

A. They vary very much with age. A weighing just made shows ten old ones weighing 13½ pounds. I have no new ones to weigh, but they would be much lighter. Ten average combs will yield from 1⅛ to 2½ pounds of wax.

Concrete (See Cement.)

Corn Flower.—Q. Do bees gather nectar from corn flowers?

A. Yes, if by corn-flower you mean the flower Centaurea Cyanus. If you mean the tassels of Indian corn, I think they get only pollen.

Cotton.—Q. There is a large amount of cotton near Phoenix. Does cotton in Arizona yield much honey?

A. Cotton is a good honey plant in the southern states, and likely, also, with you.

Cottonwood.—Q. Is cottonwood lumber good for beehives? Is basswood?

A. Both are bad for lumber for hives.

Q. Do bees gather much honey from the blossom of the cottonwood tree?

A. I think not; if your cottonwood is like the cottonwood of Illinois.

Covers.—Q. What kind of a cover do you use?

A. A flat cover with a dead air-space covered with zinc or tin. The upper and the lower parts are each of three-eighths inch stuff, with the grain running in opposite directions, separated by strips or cleats three-eighths inch thick.

Q. Would you advise deep or shallow covers?

A. For my own use I prefer the flat cover (I have no trouble with rain beating in,) although some good beekeepers prefer deep covers.

Q. I have a lot of telescope covers 11 inches deep. Will it be all right to put them on in winter, or will they keep the bees too warm?

A. No danger of keeping too warm.

Q. What do you think of the "Colorado" cover?

A. It's a good cover.

Q. Are metal-roof covers for hives with inner covers better than wooden covers? If so, why?

A. Their chief advantage is that they are always rain-proof.

Cow Peas.—Q. Do bees gather honey from cow peas? We had about three acres of cow peas here last year, and it appeared that all our bees worked on them for three of four weeks, as it seemed there were thousands, and the queer thing to me was that they did not work on the bloom, but on the joint just below the bloom or young pea. Was it wax or honey?

A. Cow peas are counted honey-plants. There are different plants which, at least at times, secrete nectar elsewhere than in the blossoms. When you see bees working as busily as you say they were on your cow peas, you may be sure they were getting either nectar or pollen. If you see no pollen on their legs you may be sure they are getting nectar. They don't gather wax, they secrete it; but they gather bee-glue.

Cucumbers.—Q. How is cucumber as a nectar-yielding plant? How many colonies could be kept at one place to the best advantage, when the farmers raise one-quarter to two acres each?

A. Hard to tell. Depends somewhat upon size of farms. If each farmer plants half an acre, you will readily see that there will be four times as much pasturage if the farms average 40 acres as if they average 160 acres. I should guess that 100 colonies might do well with one acre in every 100 in cucumbers.

Cushions, Chaff.—Q. What is the best way to make chaff cushions for hives to winter bees in?

A. Make a plain bag a little larger than the size of the required cushion closed on all sides except enough for an opening on one side to admit "stuffing." At each corner sew a straight seam as long as the depth of the cushion. Don't sew it with the bag lying flat for that would spoil the shape of your cushion. Instead of that pinch the cloth together sidewise at each corner, making a seam that will be vertical in the finished cushion, making the cushion box-shaped. Fill the cushion and sew up the hole. Cork dust may be used as stuffing instead of chaff.

Cypress.—Q. What do you think of cypress hives? Are they as good as white pine?

A. Probably they are as good; some say they are more durable.

Cyprians.—Q. Are the Cyprians better honey gatherers than the Italians? Of what color are they?

A. They are not generally considered better. Cyprian bees look very much like Italians, but the yellow bands are a trifle wider and deeper in color, more like copper.

Q. Are Cyprian queens more prolific than other races?

A. I don't think they have that reputation.

Q. Have you had any personal experience with Cyprians? If so, describe the hustling qualities, comb-capping, comparative size of bee, color (full), longevity, and disposition of the pure bee.

A. I never had but one colony of Cyprians, and that was several years ago, and I can only tell about them as I remember them. In industry, comb-capping and size, they did not especially differ from Italians, if at all. I do not recall whether they differed in color, and I know nothing about their longevity. They have the reputation of being very cross, but did not distinguish themselves in that way. The most notable thing about them was that they would start the largest number of queen-cells of all the bees I ever knew.

Daisies.—Q. Do wild daisies produce nectar?
A. No, not to speak of.

Dampness (See Moisture.)

Dandelions.—Q. Do bees gather honey from dandelion blossoms?

A. Yes, they gather a large amount from dandelions. It comes rather early for surplus, but it is of immense value for brood-rearing. Dandelions also have a great deal of pollen, which helps in early brood-rearing.

Danzenbaker Hive.—Q. Is the Danzenbaker hive as good as any for comb honey?

A. Some are enthusiastic over it; some condemn it severely. After a limited experience with it, I still prefer the regular 8-

frame dovetailed hive. A single objection would bar it out for your use. I had more pollen in sections with one Danzenbaker hive than with 50 others, probably because of its shallowness.

Dead Colonies.—Q. What is the best thing to do with hives in which bees have died during the winter? There is quite a lot of honey in them.

A. They're the nicest sort of things in which to hive your swarms. Keep them shut until you need them, to keep out robber bees and moths.

Decoy Hives.—Q. Will you please explain decoy hives. I have

Fig. 14.—Decoy hives on the roof of a shed-apiary.

seen the word used several times in the American Bee Journal. I believe that they are used to attract swarms.

A. Leave an empty hive anywhere where a swarm may enter of its own accord—that's a decoy hive.

Q. How do you fix decoy hives to catch swarms?

A. There is no fixing needed, any more than in getting a hive ready for a swarm. If you put in the hive one or more empty brood-combs it will be more attractive to the beemoth, for which you must look out.

Q. In the decoy hives will strips of foundation in the frames do as well as frames of comb? Will the bees take to the foundation as readily?

A. No; old combs are away ahead of foundation; indeed, I suspect an entirely empty hive is nearly as good as foundation.

Desertion of Swarms.—Q. A neighbor of mine says that when he kept bees and was ready to hive a swarm, he would first wash the hive thoroughly with salt water, and then hive the bees; and said he never had a swarm leave when he hived it in that way. What do you think of it?

A. Washing out a hive with salt and water is an excellent thing, if the hive is dirty. It might do as well without the salt. If the hive is clean, it may do as well without any washing. The principal precaution against having a swarm desert the hive is to see that the hive is well shaded and ventilated. You can wash a hive in an ocean of salt water, and if you set it in the hot sun with a small entrance, a swarm may desert it.

Q. I had 32 colonies of bees, and I have lost five of them. They will swarm and come out of their own hive and settle on the outside of some of the other hives, and leave their own hives empty, with lots of honey in them. When they settle on the other hives it causes a fight. What makes the bees do this?

A. Bees sometimes seem to have a mania for deserting their hives in spring and trying to force their way into other hives, and it isn't easy to say just why. Some think because they are weak and discouraged. Some think because they have started a lot of brood, and then the old bees have died off so rapidly that enough are not left to cover the brood. In any case the advice given is to have only strong colonies in the fall. This is sound advice on general principles, even if there should be some absconding the following spring in spite of strong colonies.

Diseases and Enemies of Bees (See Foulbrood, Dysentery, Bee Paralysis, Moths, Isle of Wight.)

Distance Bees Fly.—Q. How far will Italian bees go for nectar in a fairly good clover location, with 100 colonies in the apiary and about 100 acres of alsike within two miles of the apiary?

A. Italian bees, or any other bees, work perhaps to good advantage a distance of one- and one-half to two miles—perhaps farther. In the cases you mention they would probably go that distance. The lay of the land governs to some extent the distance of their flight.

Q. My apiary is 1⅛ miles from the Red River bottom—a bottom about eight miles wide, containing a very dense forest. It is about five miles to the river where there is a very extensive agricultural business carried on. I can see my bees going to the bottoms. How far do you think they will go in the bottoms?

A. Bees have been known to go as much as seven miles, but probably not with profit more than two or three.

Q. How far will a drone fly from a hive? How far will a virgin queen fly from a hive?

A. I don't know, and I'm afraid you'll never know. I think it has been said that a drone may meet a virgin whose home is four or five miles from his home, although as a rule such long flights are not made. Some think that a mile is as far apart as the two homes usually are. But if you knew exactly how far apart the two homes are, you are still in the dark as to how much of the distance is made by the drone and how much by the virgin.

Dividing (See Increase.)

Division-Boards.—Q. Of what use are division-boards, and how often should they be used?

A. A division-board, properly so-called, is a thin board more or less tight-fitting that divides a hive into two separate compartments, as when a hive is to be used for two or more nuclei, or when a colony is too small to occupy the whole of the hive. In this sense there are very few division-boards, but a dummy is really the thing that is meant. A dummy is loose-fitting, not longer nor deeper than the frame of the hive. It may be less than that. Dummies are in use in my hives all the time, winter and summer. The frames do not entirely fill the hive, and the dummy fills up the vacant space at one side. It is much easier to get out the dummy than to get out the first frame where there is no dummy, and after the dummy is out it is easy to get out the frames. If less than the full number of frames is in the hive, one or more dummies are placed next to the exposed frame.

Doolittle System of Honey Production.—Q. What is the Doolittle system of comb-honey production?

A. A book called "A Year's Work in an Out Apiary" gives in full the system that Mr. Doolittle follows, which is a combination of good things more or less in general use, given by the author in an interesting way. Of course, it would be out of the question to give details here, but only one special feature may be mentioned, and that is that early in the season he puts over the hive a second story containing combs with more or less honey, an excluder between the two stories, and then when the time comes that there is danger of swarming, or just before the honey-flow, he takes away the brood of the lower story, giving the colony the combs of the upper story.

Drone-Brood.—Q. Is it common to find considerable drone-brood in worker-cells in colonies where all combs were drawn from worker foundation, the drone-brood being started in spring at the beginning of brood-rearing, and a considerable quantity of it being intermixed with the worker-brood?

I notice it in one of my two colonies, and it seems to be largely in the upper part of the comb. Is it on account of the foundation sagging, thus making the cells a trifle larger? The queen in this hive was of last season's rearing, would you think because of this drone-brood that she was inferior?

A. It is not common.

If the cells in the upper part of the comb are larger because of stretching of foundation it may have some effect in preventing the

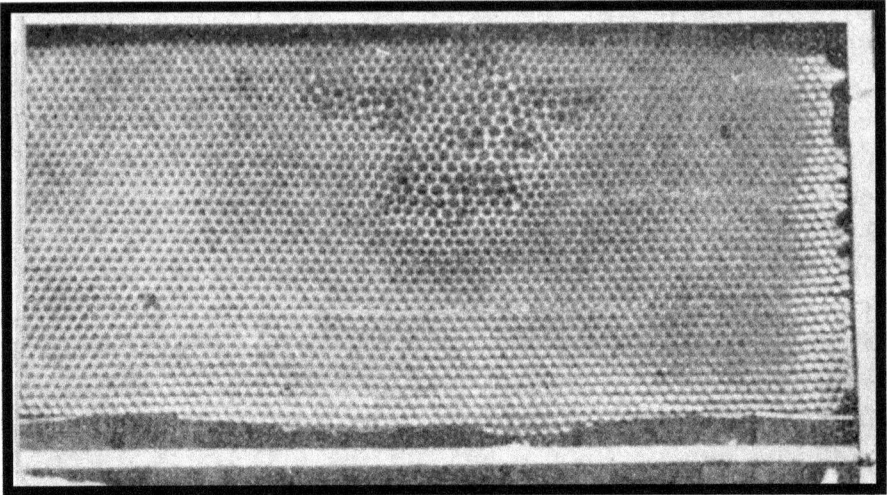

FIG. 15.—The bees patch up holes in combs with drone-cells unless worker-comb or foundation is inserted by the beekeeper.

queen from laying in these cells, and if she does lay in them the eggs may be drone-eggs. If drone-brood is found only in these enlarged cells, it ought hardly to condemn the queen. If, however, drone-brood is mixed in with the worker-brood of regular size, the probability is that the queen is beginning to fail, no matter what her age and very likely it will not be long till she becomes a drone-layer.

Q. Is it safe to uncap drone-brood and then put it back in the hive for the bees to clean the cells?

A. Entirely safe; but you can save the bees the labor of cleaning out the cells, and also save the considerable amount of food fed to the larvæ if you cut out each patch of drone-comb and put in its place a patch of worker-comb.

Drone-Comb.—Q. In reading the American Bee Journal, I see drone-comb spoken of a great deal. Please explain how I may be able to know drone-comb.

A. Lay a rule on the surface of the comb. If the cells measure five to the inch they are worker-cells. If they measure four to the inch, they are drone-cells.

Q. How many drone-cells do you think it is necessary to have in each hive? Some of my hives had so much drone-comb that I melted some of it up and put in foundation.

A. I doubt that it is really necessary to have any drone-comb except in those hives which contain such good stock that

Fig. 16. Even with full sheets of foundation, many times more than enough drone-brood will be built in odd places by the bees.

you want to rear drones from them. Some, however, believe in having perhaps two inches square in each hive.

Q. If bees are given full sheets of foundation will there be any drone-cells?

A. None to speak of; but the bees are likely to find some little vacancy that they can fill with drone-cells. Practically speaking, however, there will be no drone-comb if the frames are entirely filled with worker-foundation.

Q. When is the best time to cut out drone-comb to prevent more being built?

A. It doesn't matter when, if you fill the hole with worker-comb or comb-foundation, only it will be easier to do it in spring, when the comb is empty.

Q. Do you think bees will rear workers if shaken in a hive with a queen, in a full set of all drone-combs?

A. I tried that once, and the bees wouldn't stay; swarmed out. In other cases, where there was an excess of drone-comb, they reared an excess of drones; but in some cases they narrowed the mouths of the drone-cells and reared workers.

Drone-Eggs.—Q. How can I get the queen that I want, to lay drone-eggs? If I give drone-comb they rear workers just the same.

A. A little before harvest time, strengthen the colony by giving it additional sealed brood from other colonies, and if there is drone-comb in the brood-nest she'll lay in it.

Drone-Layers.—Q. Does an old queen ever get so she will lay only drone-eggs?

A. In many cases the contents of the spermatheca become exhausted, which will be shown by part of the brood hatching out of worker-cells as drones, finally there being only drones.

Drone-Trap.—Q. I have a drone-trap or swarm-guard. I don't have any success with it. How should I use it, and why should I catch the drones?

A. A drone-trap attached to the entrance catches the drones as they attempt to leave the hive, when you can maltreat them in any way you wish. The intention generally is thus to suppress the drones of the poorer colonies, leaving the chances in favor of having your virgin queens fertilized by drones from your best colonies. In the same way you may catch the queen of an issuing swarm, should one issue when you are not present, thus preventing the swarm from going off with the queen, and allowing you to remove the brood and leave the swarm with the queen. But this does not settle matters, for the bees may go on swarming so long as the queen is with them, and when a young queen emerges from her cell the bees will swarm again, and if the young queen is prevented from going out with a swarm she will also be prevented from going out to be fertilized, and then, if she lays at all, she will be a drone-layer.

Drone-traps should be used only in extraordinary circumstances, and are rarely used by practical beekeepers.

Drones.—Q. At what time do bees begin to rear drones?

A. Eggs are likely to be laid in drone-cells as soon as there is a considerable flow, and drones will appear 24 days later.

Q. Will drones stay with a colony of bees without a queen?

A. Yes, better than with a queen.

Q. Two of my hives have some drones yet (November.) Why is it?

A. I'm afraid they're queenless; yet it sometimes happens that drones are suffered late where there is a good queen and plenty of honey.

Q. I find that one of my colonies is still rearing drones. The queen looks all right. She has been one of the best among 65. She is supposed to be young as she came through the mail in May, and I started her with a small bunch of bees, and she built up a strong colony. I didn't notice any drones until lately. (January.)

A. The queen may be all right and she may be all wrong. It sometimes happens that a colony takes a notion to cherish some drones after drones are generally killed off, keeping them through the winter, while the queen is all right, but the fear is that your queen has become a drone-layer, even if she is not old. You can probably tell by the sealed brood next spring or even now, if there is any sealed brood present. If you find cappings of worker-cells flat, that's all right. If they are raised and rounded, like so many little marbles, the queen is a drone-layer, and should be killed. To be sure, there has been known such a thing as a queen getting over being a drone-layer, as W. M. Whitney has reported, but you better not count on that.

Q. My bees had no drones to speak of this season, except on two or three days, when I saw four or five flying from two hives, and the bees killed them right away. What was the cause?

A. The absence of drones may be due to the poorness of the season. Keeping drones is a sort of luxury that bees indulge in when they are prosperous, and when forage is scarce they do not feel they can afford it.

Q. My nearest beekeeping neighbor is a mile and one-quarter. If I stock up with Italians is there much danger of my queens being fertilized by his black drones? I use full sheets of foundation, and have very few drones. He uses only starters, and I saw whole frames in his hives that were built out solid with drone-comb, except two inches where the starter was. He had six colonies, and got no surplus. They swarmed as soon as they got a half-gallon of bees in a hive, and I don't want any of his stock, but would like to rear most of my own queens. Two of those I reared were larger and better layers than the one I bought.

A. The probability is that your neighbor's drones will be obliging enough to meet most of your queens. Can't you get him to change to Italian blood?

Q. Would you advise rearing drones and queens from the same mother?

A. It will be better to rear queens from your very best colony and drones from a few of the next best. Yet if you should try to rear queens and drones both from the same colony it is not certain that much harm would come from it, for the young queens would be likely to meet drones from other colonies, perhaps from a colony a mile or more away.

Q. I thought I saw a few black drones in an Italian colony. Do you think I was right, or was I fooled in the kind of bees?

A. Nothing strange about it. Drones are freebooters, and in prosperous times will be accepted in any colony. So black drones may have come from some other colony. It is also true that pure Italian drones are sometimes very dark when the workers are properly marked.

Q. Are the drones from a mismated Italian queen still pure Italian, or are they hybrids?

A. It is generally considered that the drone progeny is not affected by the mating of the queen, although some maintain that the blood of the queen may be so affected as to affect the character of the drone progeny slightly in the direction of the drone which the queen met.

Q. I have one colony that has two kinds of drones. About half show yellow bands, while the others do not. The workers do not all show three yellow bands. What race are they?

A. The drones are not uniform, and only the workers are relied on to decide purity. Your colony of bees that do not all show three yellow bands is hybrid unless some bees have entered from other hives—a thing that often occurs. To be sure entirely, examine the young bees that have not left the hive; if all of these have three yellow bands you may count them Italians.

Q. About how many drones should there be in a healthy colony?

A. Some think it best to try to keep them down altogether, except in one or more of the best colonies. I think G. M. Doolittle allows to each colony what drones they can rear in a square inch of drone-comb.

Q. When should the drones be caught? Why are there so many when it is only necessary for them to meet the queen once?

A. Any time. Prevention, however, is better than cure. Allow very little drone-comb in your hives and you'll have few drones.

For greater safety to the queen. If there was only one drone for each queen, the queen might make many trips before mating.

Drumming.—Q. What do you mean, in answering queries, by "drumming" the bees out of a hive in transferring, and how is it done? Is it knocking on the sides or top, and for how long, and how hard? Do you use just the fingers, or fist, or stick?

A. Turn a hive upside down, drum on the sides of the hive with your fists or a heavy stick on the opposite sides, and if you drum long enough with heavy strokes you will set the bees to running up into whatever is placed over. No light tapping with your fingers will do, neither with the fists, unless strong and heavy.

Dummies.—Q. What are dummies? What is their construction and use?

A. Take a top-bar and nail a board on so that the length of the board is the same as the length of a frame, and the depth of top-bar and all the same as the depth of frame, top-bar and all. That's your dummy. It may be an inch thick, or anything less down to one-quarter inch. It is used to fill up any space desired, and especially at one side of a hive. If no dummy is in the hive it is hard to get out the first frame, if the frames are self-spacing or fixed-distance frames. If there is a space filled with a dummy at one side, it is easy to take out the dummy, and then easy to take out any desired frame.

(See also Division-Boards.)

Q. You are often called upon to explain what dummies are, how they are made, and how used in the hives. In confining a small colony to one side of the hive, do you fill the empty space with anything?

A. A weak colony, say one that needs only four frames, may have a dummy at one side of the frames with the remaining space in the hive left entirely vacant; only the dummy must be moved and a frame or frames added as needed. Generally, however, when one has a weak colony of that kind which is expected to build up, one has enough empty combs to fill up the hive, and in that case I wouldn't use a dummy at all. You may ask whether the bees would not be warmer to have the combs that are occupied shut off from the empty combs by a dummy. One would naturally think so, yet experiments carefully made, if I remember rightly, by Prof. Gaston Bonnier, showed that the empty combs were just as good as a board partition.

For winter, the space behind a dummy may be filled with warm absorbents.

Dwindling.—Q. (a) Why do some colonies (having plenty of stores and a fairly good number of bees) start brood-rearing in the latter part of winter and get a good deal of capped brood and brood in all stages, and when cold weather comes the whole outfit dies? This has happened with me two seasons.

(b) How can I avoid this thing?

A. (a) This seems to be a case of what is called spring dwindling. The cause is somewhat in doubt. It looks a little as if the bees were old, had more brood started than they could take care of, then died off with the strain of trying to provide digested food for the brood, sometimes swarming out with plenty of food in the hive.

(b) I don't know, unless it be to have colonies strong with bees not too old the preceding fall.

Dysentery (See Diarrhea.)

Dzierzon Theory.—Q. The following was copied from a daily paper. Is the doctrine true? I have never heard of it before.

"The strangest thing that Mr. Watts told the Review reporter was that the drones are produced from unfertilized eggs. One with experience with poultry would expect such eggs to fail to hatch. Scientists, both by microscopical examination of the eggs found in drone combs and by studying the life history of bees, have proven that the drone actually has only one parent, the queen mother, and every observing apiarist has seen convincing evidence of this fact."

A. Of all the bee journals of any language in the world, the one that I have valued most is the first volume of the American Bee Journal. That was published in 1861. Its chief value consists in the fact that it gives a full discussion of the Dzierzon theory, the kernel of which is that the queen is fertilized once for life, laying fertilized and unfertilized eggs, and that the unfertilized eggs produce only drones. In the half century since then there has been some attempt to controvert the Dzierzon theory, especially by Ferdinand Dickel, but intelligent beekeepers quite generally accept it; so that the clipping is all right.

Egg-Laying.—Q. When does the queen begin laying in the spring?

A. In a colony wintered outdoors she begins, in the north, in February, or even in January. In Texas, probably earlier. If cellared, she begins about the time bees are taken out of the cellar.

Q. What would you think of a queen that fills every cell in most of the combs with eggs, and in numerous places has eggs in half-built cells, and in cells filled with beebread?

A. That is just what every good queen should do, except laying in a cell containing pollen. When you find eggs in a pollen-cell you may generally count that laying workers are present, although it is possible that occasionally an otherwise good queen may do such a foolish thing.

Eggs.—Q. Has the queen the power to fertilize eggs or not?

A. Sure. She fertilizes all but the drone-eggs.

Q. In regard to bee-eggs, is there any difference or distinction between the eggs from which a queen and worker are hatched or reared? If I am correct, bee-men use any egg they may come to when transferring eggs to queen-cells, and the difference results from the size of cell and the material on which the young bees are fed.

A. An egg laid by a good queen in a queen-cell is precisely the same as one she lays in a worker-cell. A drone-egg is a different thing. A drone-egg is unfertilized and can produce nothing but a drone, even if fed in a queen-cell; other eggs are fertilized.

Q. I have only one colony of bees, in which I find many cells with from 2 to 6 eggs in each. And at the front end of some of the combs there are cells that seem to have 30 or 40 eggs in each. I never saw anything like it before. I could not find the queen. Did laying workers try to fill the cells with eggs?

A. Almost certainly it is laying workers. You will probably find that if any drone-cells are in the brood-nest the nuisances have been specially favorable to them. Also, you will be likely to find one or more queen-cells, and in these there may be as many as a dozen eggs in each. Better break up the whole business, giving combs with adhering bees to other colonies.

Q. I have a queen that I reared in a nucleus. She is of good size and pure Italian; very gentle. I have seen her lay while holding up the comb, but I have counted as many as six eggs in one cell. What do you think is the matter with her? She is in a hive, but the bees cover only four frames in it. Do you think there ought to be more bees in it so the queen could have more room?

A. It is nothing unusual for a good queen to lay more than one egg in a cell when she has so small a force of bees that she hasn't room to spread herself; although it is unusual for her to

lay so many as six in a cell. If she keeps supplied with eggs, all
the cells that the bees cover, you needn't worry about her throwing in a few for good measure. If, however, she lays duplicates in
a few of the cells and leaves other available cells empty, there is
something wrong, and if she persists in that line of conduct she
should lose her head. But it happens sometimes that a queen
will lay in an abnormal manner for a week or so, and then
straighten up and lay as good queens should.

The likelihood is that your queen is extra good.

Q. Why do queenless colonies eat or destroy eggs given them
to rear a queen? One of my colonies destroyed a cell I gave it,
and is queenless yet.

A. Bees frequently eat or destroy eggs given them or left
with them when queenless. I don't know why. They will also destroy queen-cells sometimes for no apparent reason.

Q. How long can combs of eggs and unsealed brood remain
off a hive without being damaged?

A. I don't know. That's a good subject for you to experiment on. I know that brood nearly ready to seal will begin to
crawl out of the cell within a few hours—perhaps two or three—
after being taken from the hive. In Switzerland they make a
practice of sending eggs by mail, so it is likely eggs will keep at
least a day or more. A fresh-laid egg would perhaps keep better
than one three days old.

Entrance-Blocks.—Q. Are entrance-blocks used on the hive
all year around? Or when would you advise me to put them on,
and what opening?

A. The entrance-blocks should be taken away entirely during
hot weather, or while in the cellar. For outdoor wintering they
should be used to make a small entrance. Then in spring enlarge
them only as the entrance becomes crowded.

Entrance-Guards.—Q. Is it dangerous to put entrance-guards
at the entrance with ventilation at the top for preventing
swarms?

A. If the opening for ventilation is large enough for bees to
pass through, entrance-guards will have no effect whatever.
Neither will entrance-guards have any effect in preventing
swarming; all they do is to catch the queen when the bees swarm.
Of course, when the queen is caught in the guard the swarm will
return; but there will be trouble later.

Q. Can a queen pass through an entrance-guard?

A. Not if the entrance-guard is perfect and the queen of normal size. Some have thought that when a queen is not laying, her abdomen consequently smaller than usual, she might get through a perforation smaller than when in full laying. But it is not the size of the abdomen that prevents her passage, it is the thorax. The abdomen is soft and yielding, and when at the largest it will easily flatten out to go through any perforation large enough to allow the passage of the thorax. The thorax is a sort of bony structure, which is the same whether the queen is laying little or much.

Entrances.—Q. I have contracted the entrances to all hives of colonies that need feeding or that are weak in bees. The strong colonies don't need any contracting, do they?

A. It is not so important to lessen the entrance, as to avoid everything that may start robbing. This year my nuclei have the same entrance as the full colonies—12 by 2 inches—and there has also been one case of robbing at a full colony with a normal laying queen. Very likely some unwise thing had been done to start the robbing.

Q. Do you contract the entrance in the spring during cool nights? If so, how much? Is it not a good plan to contract the entrance on account of robber-bees in spring?

A. Yes, just as soon as my bees are taken out of the cellar the entrances are contracted to a hole three-quarters to one inch square. It helps against robbing and keeps the bees warmer, day and night.

.Q. Would you contract a wide entrance during a cool spell in summer?

A. No. Takes too much work. But if I had only a few colonies, and worked them as a sort of pastime, I might change the entrance according to the weather.

Q. Is there any advantage to have the entrance 1½ inches deep and full width of hive?

A. Yes; it gives chance for better ventilation in hot weather, and also in winter, if you winter in the cellar. But you cannot have 1½ inches under bottom-bars in summer unless you have some provision to prevent the bees building down.

Q. Should I diminish the entrance of the hive in winter? (California.)

A. In your locality probably no contraction is needed.

Q. Should the entrance be 1x5 inches, with a wire-cloth in it to prevent mice entering?

A. Wire-cloth with three meshes to the inch is a good thing at the entrance for winter, but not when bees are flying daily.

Q. If a colony is extra large, how large should the hive-entrance be?

A. Full width of the hive.

Q. What do you think of having the entrance the long way of the hive seven-eighths of an inch high during the honey flow? Did you ever try it?

A. If you mean to have the entrance the long way of the hive, and that the only entrance, I shouldn't like it so well as to have the entrance the usual way, because the latter allows freer entrance of air. In Europe it is quite common to have the entrance as you describe. That's called the "warm arrangement," and the frames running at right angles to the entrance (the common way here) is called the "cold arrangement." I never tried the single entrance at the side, but have practiced largely having the entrance on all four sides. I like it much, but now have only one opening two inches deep, as being, in the long run, more convenient.

Q. How about a separate entrance to supers?

A. Some advise it, but generally it is not used. An opening above for ventilation, in very hot weather, however, may be a fine thing.

Q. Does it make any difference if the hive-entrance faces the north during the winter, and would it be a good plan to build a sort of box around and close the entrance one-half its present width? (Ohio.)

A. So far south as southern Ohio it probably makes little difference how a hive faces. Yet a good many favor facing south, and having no protection on the front. In this way the bees more quickly get the effects of the sun on a warm day in winter.

Equalizing Brood.—Q. Is it a good policy to equalize brood in the spring?

A. Yes, if rightly done, and no brood taken from any colony unless it has more than four frames well filled with brood.

Excluders.—Q. Do you use a queen-excluder on your hives to keep the queen from laying in the sections? If not, how do you prevent this?

A. With full sheets of foundation in sections, and frames not too shallow in the brood-chamber, the queen so seldom makes trouble in the supers that I never use an excluder to keep her down.

Q. Are not queen-excluders a hindrance to bees, or will I have to get some excluders? If so, how many?

A. While it is generally thought best to use excluders for extracted honey, some do not use them, such prominent men as C. P. Dadant and E. D. Townsend being of the number. The latter says that by giving additional supers always on top he has no need for excluders. If you find it is better to use them, you will need one for each colony.

Q. Have you tried the new queen and drone-excluders, or honey-boards, made of wire? Have they any claim to be classed as an improvement on the perforated zinc, or is it only a scheme of the manufacturers?

A. I do not use excluders under supers, so I don't use many excluders, although for some purposes they are indispensable. Having quite a stock of the old kind of excluders on hand, I have never tried the wire excluders. I don't suppose there is a great deal of difference, but one would suppose that the bees would like the smooth wires better than the sharp edge left by the punching of the metal for the perforations.

The wire excluders also allow better ventilation.

Q. Can virgin or unfertile queens pass through excluding zinc?

A. A laying queen looks much larger than a virgin, but it is the abdomen that's larger, not the thorax. It's not the abdomen, but the thorax that prevents a queen going through the zinc, and I think the thorax of a laying queen is no larger than it was when she was a virgin; so she ought to go through no more easily one time than another. But a virgin queen probably makes a more vigorous effort to go through, so she might go through an aperture through which she would not force herself after she settled down as a laying queen.

Extracted Honey.—Q. I am going to buy five dovetailed 10-frame hives this spring. I only want honey for the house. Which is better for me, the extracting hive or sections? I read in the bee-books the extracting hive is best for home use. Please tell me why.

A. Extracting saves the bees much labor in building comb, so it is generally estimated that you can get about one-half more extracted honey than comb. So, in deciding the question for yourself, the question is whether you would rather have 100 pounds of comb honey or 150 pounds of extracted.

Q. Is it advisable to extract honey as soon is it is gathered? Is there any danger of it getting sour?

A. It may be extracted early if it is sealed; otherwise not till the crop is well over. It may sour, and the flavor may be poor.

Q. Owing to the lack of supers I extracted some honey when about two-thirds capped. Will it do to sell it that way?·

A. If the honey is very thin, it is better not to sell it in that condition, but the mere fact that· a third of it is still uncapped does not condemn it. If it is good, thick honey, it does not matter that it was partly unsealed. If thin, it may be brought to a better consistency by letting it stand uncovered where it will be heated to 100 or 125 degrees.

Extracting-Combs.—Q. Can I get honey out of the extracting-frames without the extractor? Can I melt it over the stove some way without breaking the comb, and will the bees store honey in the comb again?

A. No; if you want to save the combs, it's the extractor or nothing.

Q. Can good extracting-combs be built in Hoffman wired brood-frames from 2-inch starters of medium brood-foundation? Will they stand extracting as well as combs built from full sheets?

A. Yes, fairly good.

No; and for two reasons. Most of the comb will be built without any foundation, and the septum of natural comb is more tender and thinner than that in foundation. Also, the wires in this natural comb will not ·be all in the septum, as will the wires in full sheets of foundation.

Q. Can new combs be used for extracting when built on full sheets of comb-foundation and wired?

A. Yes; but while they are new and tender it is well to use caution in extracting, if they are very full. Turn not too rapidly, and extract perhaps half the honey on one side. Reverse the comb and extract all the other side. Then reverse again and finish the first side.

Q. Which would you recommend, the 8 or 10-frame, full or shallow super, for the production of extracted honey?

A. Ten-frame hives, or larger, for brood-chamber, and I think I should prefer shallow extracting combs.

Extractors.—Q. Will it pay me to get an extractor for twenty colonies?

A. Yes; or for three, especially if you expect to increase.

Q. I am thinking of buying an extractor. What kind would you advise me to get? How about the Novice 4-frame nonreversible extractor? Is the Cowan rapid-reversible any better? Is the 4-frame too big, or not? Does the reversing help any?

A. It is generally well to make sure that your extractor is too large rather than too small, taking into account the possibility of increasing the number of your colonies. So, if you don't object to the difference in price it may be well to get the 4-frame. The reversing is a decided advantage, although one kind does as good work as the other.

Q. Would like to know the speed at which a honey extractor must run to do good work. I have some cogwheels speeded three turns of the smaller to one of the larger. Will that speed enough to extract honey?

A. Three to one will give you plenty of speed. Indeed, there is no trouble about getting speed enough with no cogs at all. The first extractor I knew anything about had none; each revolution with the handle made a revolution with the baskets.

Q. How long may an extractor remain without washing? That is, how long may the extractings be apart without injuring anything?

A. I think in some cases harm might be done by leaving an extractor daubed for 24 hours. I know that in some cases a week or more will do no harm. Perhaps the kind of honey or the condition of the atmosphere makes a difference.

Tin does not readily blacken the honey, but all iron parts do. Better wash the extractor often.

Eyes of Bees.—Q. How many eyes has a honeybee?

A. They don't all have the same number. For the sake of making the count easier, we may say the worker has three simple and two compound eyes, each of the compound eyes being made up of a number of facets; but really each facet is a separate eye. Cowan says: "There is great variation in the number of facets in the compound eyes of bees. In the worker the lowest is given as 3,500, whereas we have ourselves found as many as 5,000." Drones have more than either queen or worker. Cheshire counted on each side of the head—in a worker, 6,300; in a queen, 4,920; in a drone, 13,090.

Fanning of Bees.—Q. In warm weather, when the bees are fanning, do they do that to get the water out of the honey, or to cool the hives?

A. Both; but perhaps more than either to get fresh air into the hive. Bees seem to have a notion that pure air is a fine thing, summer or winter.

Q. When the bees are vigorously fanning with their heads to the entrance, which is accomplished, cool air driven in, or hot air being drawn out?

A. When I have put the back of my hand near the entrance it has always felt as if the current were toward my hand, and so drawn out of the hive.

Feeders.—Q. What is the best feeder to use for any amount of feed?

A. If a considerable amount of feed is to be given, nothing is better than the Miller feeder. The Doolittle is excellent for smaller amounts and handy for the bees. For an entrance-feeder the Boardman is good.

Q. If I have a correct idea of the Alexander feeder, it is used under the bottom-board of the hive. How would the bees get access to the feed?

A. The feeder is, so to speak, part of the bottom-board, at the back end of the hive, on the plan of the simplicity feeder, so the bees come directly down from the frames into the feeder.

Q. What do you think of the Boardman feeder?

A. Good; but when heavy feeding is to be done you would expect me to prefer the Miller.

Q. Would you recommend the division-board feeders for beginners?

A. They are excellent where you do not care to feed a larger amount than they contain.

Q. I wish to feed some colonies I have bought which are light in stores. In using the Doolittle feeder where, in the hive shall I put it—as an outside frame, or in the center of the hive?

A. Don't think of dividing the brood-nest, but put it next to the first frame that contains brood at one side.

Feeding Back.—Q. Do you endorse the suggestion of Alexander as to running part of one's colonies for extracted honey and feeding back into the comb-honey hives to provide continuous supplies there for night work, and at times when weather prevents field work?

A. I never made a success of feeding back to have the honey filled in sections.

Feeding and Feeds.—Q. I had one colony and lost it by feeding them only sugar water. Other bees robbed them and they starved. What is the best feed, and how and when shall I feed them?

A. The best thing is to give them combs of sealed honey, but it isn't likely you have them. The next best is a syrup of granulated sugar, probably just what you did feed them, only there was probably something wrong about the way you fed that started robbing. Of course, I cannot tell what it was that was wrong; possibly you may have spilled some of the feed, or done something else that was a bit careless. Be careful not to leave any cracks open that will let bees in from the outside. If there is danger of robbing, it is well to give feed in the evening after bees have stopped flying, and to give no more at a time than they will clean up by morning. For fall feeding nothing is better than a Miller feeder. If you feed early, equal parts of sugar and water will be all right; but if you do not feed until after the middle of October, then you can have 5 parts of sugar (either by weight or measure), 2 parts water.

Evidently you have no bee-book of instructions, and it will be big money in your pocket if you get a good one, say such a one as Dadant's Langstroth.

Q. When is the best time to feed the bees?

A. The best thing is never to feed them, but let them gather their own stores. But if the season is a failure, as it is some years in most places, then you must feed. The best time for that is just as soon as you know they will need feeding for winter; say in August or September. October does very well, however, and even if you haven't fed until December, better feed then than to let the bees starve.

Q. Two years ago I bought two colonies of bees, and the first year they increased to five colonies. I lost one colony the spring of 1905, and last fall I had six put away in good condition with plenty of honey for winter. I just now lost one colony. I examined the hive and found the honey somewhat watery, running a little out of the hive. What is the cause of this? Can I feed the honey if other bees clean out the comb?

A. If you had examined closely you might have found that it was mostly water that was running out of the hive. Water may be found running out of a hive containing a colony in good condition, the vapor from the bees settling on the cold walls of the hive as water, and running out of the entrance. It may also settle on the unsealed honey in the combs, making the honey thin, sometimes so thin as to run out. There is nothing unusual in all this, and you need not fear to feed this honey to the bees when the weather gets warm.

This thin honey will not do for winter feed.

Q. When you wish the bees to replenish the brood-chamber, how do you feed, and where do you place the food?

A. If feed is needed in the brood-chamber, you may count on the bees putting it there in preference to any other place, no matter how you feed nor where you place the food. I use Miller feeders, placing the food on top. . The crock-and-plate plan is also good.

Q. Can I safely save scorched candy until next summer and feed it without danger to the bees—let them store it?

A. Save your scorched feed till next spring, not for the bees to store, but for them to use up in rearing brood.

Q. If in your judgment it would pay to feed bees right along through the season all the sugar at 5 cents per pound that they will use to have them make honey to sell at 15 cents per pound, will they neglect the fields to feed on the syrup?

A. It would be very unadvisable, unless you want to get Uncle Samuel after you. To feed sugar so as to sell the resulting product as honey would be rank adulteration, for the product would not be legal honey. Indeed, one should strive to avoid as much as possible feeding sugar syrup for the use of bees, lest some of it should get into the surplus. Besides, it would not pay, as so much of it is used in comb-building.

Feeding Frames of Honey.—Q. I have a lot of frames full of honey nicely capped and in a cool room where the temperature goes down to zero. I presume this honey is granulated. I intend to take those frames in the spring and divide them among my colonies as feed. Is this frozen honey good? Can the bees thaw that out, or will they carry the sugar out instead of using it for brood-rearing?

A. The honey is entirely wholesome, but very likely the bees will waste a good deal of it by carrying out the undissolved granules. You can do something to prevent that if you will go to the trouble of spraying the combs with warm water by means of an atomizer, first uncapping any cells of honey that may be sealed. When the combs are cleaned off dry by the bees they may be sprayed again. Don't begin this until the bees are flying freely.

Feeding Bees in Box-Hives.—Q. Would it do to take some of the box-hive colonies that are in danger of starving into a warm room this winter and transfer them to good frame hives, using only the good combs, and contract to the size bees will occupy, placing candy between the frames or on top? Or would it cause the bees to be over-excited, filling themselves, and when again confined in the cellar without a cleansing flight, to become filthy and sick?

A. Don't transfer in winter. Those box-hives most likely have no bottoms; if they have bottoms pry them off. Turn the hives upside down, put candy between or on top of the combs, and leave them upside down as long as in the cellar. When I had box-hives I wintered them upside down in the cellar.

Feeding in Cellar.—Q. What time can bees be fed that are wintered in the cellar?

A. Any time rather than have them die; but the feeding should all be done before putting in cellar.

Q. Is there any possible way of feeding bees in a cellar? I think some of my colonies are too short of stores for winter.

My cellar is rather warm this year on account of a new furnace. I have a separate apartment for the bees with plenty of fresh air, but it is still too warm at this date; the temperature keeps up to 55 and 65 degrees. The bees are very quiet yet.

I thought of giving each colony syrup separately in a sort of little tray so arranged that the bees could not drown. Would the bees come to get this syrup, or could it be given in some other way? How and when could it be done so that half of the bees would not rush out of the hives? My hives are put in two rows, one on top of the other, and all of the covers are off.

A. With a big lot of fresh air for the bees you will likely find that they will winter well at 55 or 60 degrees, although they will consume more stores than at a lower temperature. Still, as you say, the increasing cold will bring down the temperature. Better not let it get below 45 degrees.

If I understand correctly, your hives are raised in front by 1-inch blocks, and that makes a space of at least an inch and a half. That allows you to put a shallow dish of feed under the frames, and if your colonies are reasonably strong they ought readily to come down to the feed at 55 or 60 degrees. If it is much colder than that and the colonies are rather weak, they will not be likely to come to feed. Instead of the proposed wire screen over the syrup you may do better to cover the syrup with cork chips. You will get these from your grocer. He gets them as packing for grapes in cold weather, and generally throws them away.

If feeding below does not prove a success, you can feed above. Edwin Bevins reports excellent success with lump sugar. Wet the lumps by sprinkling water upon them, but do not make them wet enough to dissolve the sugar. Then lay the lumps directly on the top-bars over the cluster of bees.

Feeding for Stimulation.—Q. I want information in regard to feeding bees in the spring, so as to stimulate brood-rearing. How shall I proceed, especially when to commence, and what precaution to use?

A. Without a good deal of experience, you may do more harm than good. Don't begin till bees fly freely; feed about half a pound diluted honey or a syrup of sugar and water, half and half.—the honey is better. Feed in the evening, for fear of robbing. Every other evening will do. It will do no good to feed when the bees can get even a moderate amount among the flowers. Bees are apt to fly out and be chilled and lost by too early stimulative feeding.

Q. I have three colonies of bees that I am afraid are short of stores, but if they should live until spring, and it gets warm enough so they can fly occasionally, would it be all right to feed them sugar syrup in feeders on top of the frames, a small amount each day, until the flowers bloom? Would it be likely to start robbing?

A. Instead of feeding a little every day, better give them a good feast, giving it to them as warm as possible, so as to get them to take enough to tide them over a considerable space of time. If you give them a little every day when they can fly only occasionally it keeps them stirred up and makes them fly out at times when they may be chilled and never get back to the hive. If the feed is given so that no robber can get to it except through the entrance of the hive, there ought not to be much danger of robbing, especially if the feed be given well on in the day.

Feeding in Fall.—Q. When is the proper time to start feeding for winter?

A. August, if they can gather nothing later. In general, just as soon as possible after it is known that feeding will be necessary. Generally it ought not to be necessary.

Q. When shall I give the bees their large feed for winter? How many pounds of sugar should I give a colony that has very little stores at the present time? (Indiana.)

A. The sooner the better. September is none too early, but in your locality there will be warm days much later.

Twenty-five pounds of sugar is none too much for a colony that has no stores. From that you must deduct for any stores they have on hand. Remember, however, that's the weight of the sugar, not sugar syrup, and the water in the syrup will, of course, be additional weight.

Q. The honey flow seems to be over here, and I have three

weak colonies with very little comb, but nice, good queens. How would you feed them so other bees would not get to the feed?

A. Use a Miller feeder in the evening after flight is over, and there will be no trouble. Other feeders can be used. If you happen to have none, you can use a crock-and-plate feeder. Take a gallon crock, or some other size, put sugar in it, and an equal measure or weight of water; lay over it a piece of heavy woolen cloth or four or five thicknesses of cheese-cloth, and on this lay a plate upside down. With one hand under the bottom and the other on top, quickly turn the whole thing upside down, and your feeder is ready. Take the cover off your hive, set over it an empty hive-body, set your feeder in it, and cover up, being sure that all is bee-tight.

Q. How will this new plan of feeding work? Place tin containers about the size of a half-pound baking powder can **cover,** containing bee candy, above the brood-frames, inside a 1-inch wooden frame to fit on the top of the hive under the cover. These tin containers set side by side just above the brood-frames, would be in the warmest part of the hive, and their candy contents would be easily accessible to the bees through the holes between these circular tin containers. This plan of feeding is easily adjustable, as a sufficient number of feed-containers can be used for either large or small swarms with no danger of feed running out to kill the bees. Tin can manufacturers can supply these at small cost.

A. This plan would work all right, I should think. In weather a bit cold the bees would not reach the candy quite so readily as if laid directly on the top-bars. Some apiarists pour candy in paper plates for feeding.

Feeding in the Open.—Q. I have some waste honey and I am feeding the bees the honey outdoors on some wide boards. Is that as good as feeding in feeders?

A. Fully as good or better, if your neighbor's bees do not get too much of it, and if you are absolutely sure the honey contains no germs of foulbrood.

Q. We are in the midst of a protracted drouth, hardly a flower to be seen. I have filled my bee-feeders with syrup made from granulated sugar and placed the feed in the yard where all the bees can help themselves. Is this method of feeding all right, or should the food be placed in the hive?

A. Feeding out in the open is a little more like having the bees gather from the fields; only if other bees are near you they will also partake of the plunder. The stronger colonies will get the lion's share, but you can make that all right by taking filled frames from the strong and giving to the weak.

Feeding in Winter.—Q. What shall I feed this winter? Can syrup be fed, or should I feed sugar candy?

A. I would rather feed syrup in winter than to let bees starve, but it is probably about twice as safe to feed candy as to feed syrup.

Q. I have a colony of fine Italian bees which have not stores enough to last them a month. I had to take it away last summer and have not as yet got it home. How can I feed it at this late day? It is in a chaff hive with extra super filled with cushions.

A. The best way is to give combs of sealed honey. Carefully take out the empty frames and put the combs of sealed honey close up to the bees, for if there is a space between the bees and the honey, and it should be quite cold for a time, the bees might starve without ever touching the honey. What's that you say? "Haven't any combs of sealed honey?" Well, that's about what I expected. But make up your mind that you'll always have them hereafter.

Well, if you haven't combs of sealed honey, maybe you have some honey in sections. You can fit some sections in a wide frame, or even a common brood-frame, by cutting away enough of the sections to make them fit in the frame. Rather an expensive way to feed; still, I've fed a good many sections in my time.

If you haven't the sections, either, you can do quite well with candy. Take best granulated sugar and stir it into a very little hot water in a dish on the stove; but whatever you do, don't let it burn, for burnt syrup is death to bees in winter. Better not set it down in the stove-hole so the fire can touch the dish, but set the dish on top of the stove. Keep trying it, and when you find a little stirred in a saucer will grain, take it off quickly and pour into dishes making cakes three-quarters of an inch to one and a quarter inches thick. Put over the frames a cake of this candy that will pretty well cover the frames; or, if cakes are small, you can use more than one. Cover this with some kind of cloth covering, and shut up snug. Toward spring you may need to repeat the dose, but if you make the cakes thick enough and large enough no more will be needed for a good while. Your extra super on top will give you a chance to put on the candy and pack it up warm.

Fences.—Q. Do you use slats or fences between sections?

A. I have used both, but now use plain wood separators one-sixteenth of an inch thick.

Flour for Pollen.—Q. Why is rye-flour put into the hives in March?

Where can I put rye-flour in the beehives?

A. Rye-flour and other meals are given to the bees as a substitute for pollen.

If you want to put it in the hive, you can sprinkle it into the cells of a comb. But it is not generally put in the hive, but outside. Put it in a shallow dish or box outside in the sun, and if the bees are in need of it they will take it from there. But if they can get plenty of natural pollen they are not likely to touch the substitute. Use old combs for bait.

Foulbrood Versus Chilled Brood.—Q. How can a person tell the difference between foulbrood and chilled brood? I can find nothing regarding chilled brood in the text-books.

A. Chilled brood doesn't string out like foulbrood.

(Foulbrood is irregular, not all the brood dying at one time. Chilled brood is all dead.—C. P. D.)

Foulbrood.—Q. I am requeening my entire apiary with Carniolan queens, as I have come to the conclusion that the most prolific bees are the most resistant to foulbrood. How about it?

A. There is a very general belief that the introduction of pure Italian blood is an important step toward the eradication of European foulbrood and some think the same of Carniolans. It may be that there is something about Italians or some other blood through which it comes to pass that if two colonies side by side are of equal energy, one of them being of pure Italian blood and the other mostly black, the one of pure Italian blood will be the more nearly immune to foulbrood. But I doubt it. I think that Italians will fight-foulbrood better than blacks, not because they are Italians, but because they are more energetic than the others. So the most energetic bees, no matter what the kind, will be the ones that will do the most toward keeping down foulbrood. I do not remember seeing prolificness claimed as a thing to help against foulbrood. Yet prolificness helps toward it in one respect, in that it helps to keep strong colonies, and it is very important with European foulbrood that colonies be strong.

Q. Is foulbrood ever found where there is no manipulation of bees?

A. Yes, indeed. Manipulation cannot produce the disease, and the right kind of manipulation does not necessarily favor its increase; but the wrong kind does; as when a comb is taken from a diseased colony and given to a healthy one. I don't mean that giving a frame of brood from one colony to another is wrong manipulation in all cases, but it is wrong where the brood is taken from a diseased colony.

Q. Is it safe to use section-boxes over again with drawn-comb and without comb, that have been on colonies that had foulbrood?

A. I should not be afraid to use them in case of European foulbrood, but with American foulbrood there might be danger.

Q. How can I tell foulbrood?

A. The chief symptom in American foulbrood is the ropy character of the dead larva; stick a toothpick into it, and when you draw it out it will string an inch or two. If European foulbrood, look for larvæ that instead of being nearly white are quite yellowish. If you write to Dr. E. F. Phillips, Department of Agriculture, Washington, D. C., he will send you, gratis, valuable printed matter about foulbrood, and also a box so that you may send sample of diseased brood for expert diagnosis.

Q. You have written several plans for curing foulbrood. Now, if half of your colonies were diseased next spring, what treatment would you choose?

A. If they had American foulbrood, I would use the McEvoy plan. If it was European, I would wait till perhaps the beginning of clover harvest, and first see that each colony to be treated was made strong by uniting or by giving frames of brood well advanced. Then I would remove the queen and give to the colony a ripe queen-cell or a virgin queen of best stock.

Q. Is there anything that could be fed to the bees to prevent foulbrood?

A. In this country drugs are generally considered of no account in foulbrood. In England it is a common thing to add naphthol beta to the bees' food, with the idea that it helps to prevent foulbrood.

Foulbrood, American.—Q. What are the chief causes of American foulbrood? I have never heard of a case in this section.

A. The chief and the only cause is the presence of a microbe, bacillus larvæ, and the disease is generally conveyed to a healthy colony by means of honey from a diseased colony. A drop of infected honey no larger than a pin-head is enough to start the destruction of an entire apiary.

Q. What is the color of American foulbrood?
A. The dead larvæ are coffee-colored.

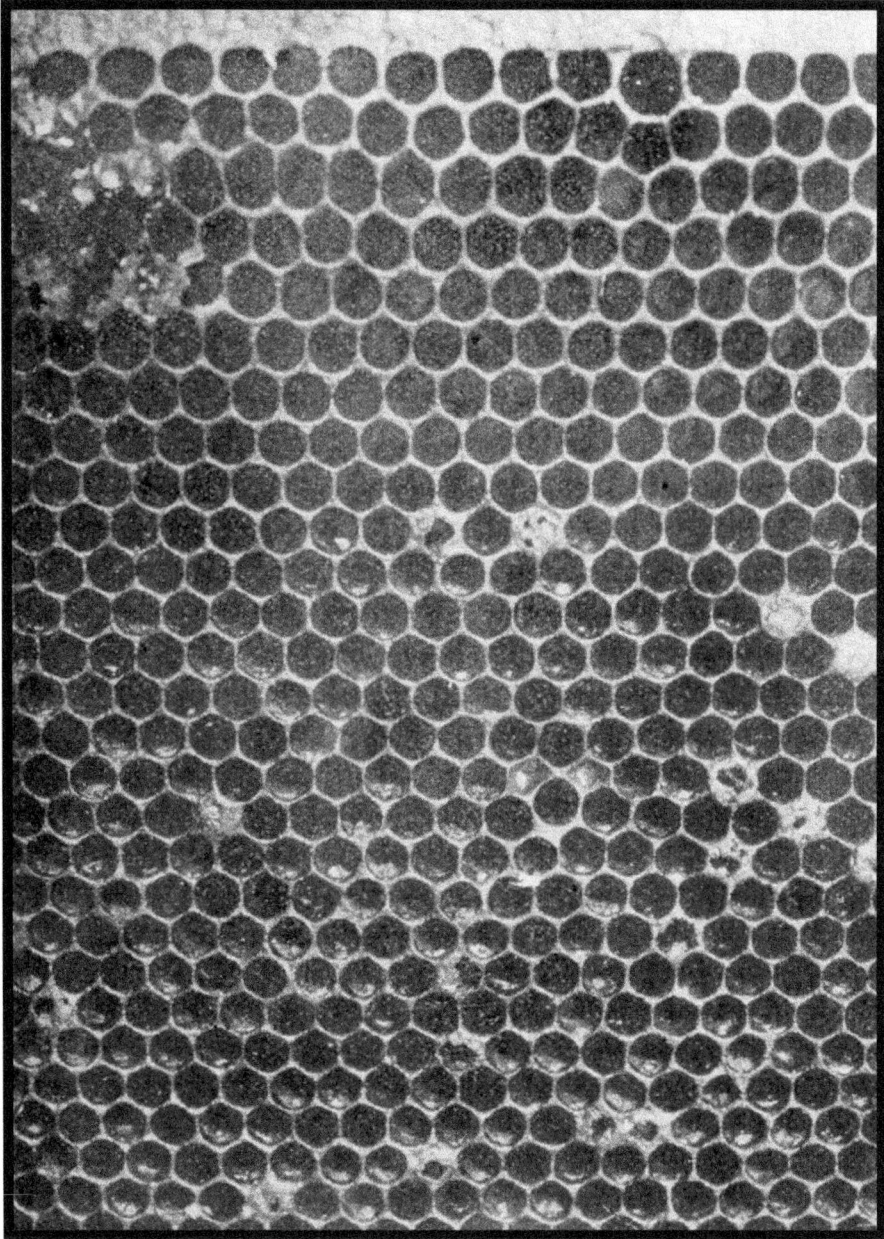

FIG. 17. American foulbrood in an advanced stage. Notice the pierced cappings and the dead larvæ in the bottoms of the cells.

Q. Does foulbrood disappear during a heavy honey-flow to show up again the following spring in the same colonies?

A. Yes, it may disappear, to all appearance, although the seeds of disease are there all the while.

Q. Can foulbrood be cured without destroying all the bees? If so, how?

A. No need to destroy the bees; the disease is only in the brood. The McEvoy plan is generally used in curing. In the honey season, when the bees are gathering freely, remove the combs in the evening and shake the bees into their own hive; give them frames with comb-foundation starters and let them build comb for four days. The bees will make the starters into comb during the four days and store the diseased honey in them which they took with them from the old comb. Then in the evening of the fourth day take out the new combs and give them comb foundation to work out, and the cure will be complete.

Q. How early in spring can bees be treated if they have foulbrood?

A. Usually no. treatment is undertaken until bees are busy gathering.

Q. How many days shall I wait after treating a colony by shaking before I can give honey or brood, if they really need it to keep from starving or dwindling on account of no young bees?

A. Perhaps five days. There ought really to be no need of feeding, for the attempt at cure should be undertaken only at a time when honey is coming in.

Q. How long is a colony immune to the disease after starting all over with fresh foundation?

A. Just as long as you are immune to the itch after being cured of that troublesome malady. In other words, if the cure of foulbrood is complete today, and tomorrow the cured bees have access to some foulbroody honey, you may count on their being diseased again.

Q. If a colony that has a few cells of American foulbrood swarms and that swarm is put into a hive containing frames with full sheets of foundation will it be in danger of having American foulbrood later on? Or is it necessary to use something like the McEvoy treatment?

A. Yes, it is in danger; but that 'later on" must not be carried too far. If the disease does not appear in the first batch of brood, you need not expect it "later on." But if there are, as you say, only a few diseased cells in the parent colony, the probability is that the swarm will be healthy.

Q. Ought I to use brood-frames which contain perfect combs,

i. e., those showing no signs of foulbrood, if purchased in a lot of hives, part of which I suspected were infected?

A. There is danger. Don't use them unless you keep a close watch.

Q. Are combs that have contained American foulbrood, and later filled with honey by a diseased colony, then extracted, safe to use again on healthy colonies over queen-excluders?

A. No. Never use again combs which have been in a colony which had American foulbrood.

Q. Are extracting-supers that have been used on hives infected with American foulbrood, after being extracted, safe to use on healthy colonies?

A. Some say yes, some say no. I suspect that the truth is that sometimes the disease is thereby conveyed, and sometimes not. It will be the safe thing to avoid using them.

Q. What is the best method to treat brood-combs so as to be doubly sure that there will be no chances of foulbrood getting into the apiary from those bought brood-combs? I have a chance to buy old combs.

A. I don't know of any way. At one time it was claimed that formaldehyde would disinfect them, but I think that is given up. Your only safe way is to buy them where you know there has been no disease.

Q. Please tell us when we shake on foundation for foulbrood whether the frames should be new, or can we cut the old comb out clean and use the frames again? I don't want to buy frames for 50 hives if it is unnecessary.

A. Generally it is considered best to burn up the old frames, but when one has so large a number as you have I think it pays to clean them up and use again. At any rate, that is what I did with quite a number. After cutting out the combs, I put the frames into a big iron kettle holding half a barrel of water into which was put two pounds of concentrated lye. The water, of course, was heated, and the frames were kept in the kettle until all wax and glue was melted off. Then the frames were rinsed in cold water to get off the lye.

Q. Is honey from a foulbroody colony fit for table use? I never heard of any foulbrood in this neighborhood and there are lots of bees here.

A. If nice and clean in appearance it is all right. Foulbroody honey that is death to bees' larvæ is entirely wholesome for human beings.

Q. After shaking one or more colonies of bees that had

American foulbrood should the smoker and all tools used be disinfected? If so, how? I put the smoker, gloves, veil, etc., in a jar and poured on them lots of gasoline, then I covered all with many sacks, weighted them down, and left them this way for one week. Do you think this will be sufficient? The gasoline was still strong and would burn vigorously after one week.

A. I don't believe gasoline kills the spores, and so I doubt its being an effective disinfectant. A solution of carbolic acid is used by some. Even carbolic acid does not destroy the spores, and I am a little bit doubtful of the need of anything more than soap and water, only so that any remains of the disease may be removed.

Q. I have 40 colonies of bees with American foulbrood. I would like to treat them in the spring. Would it be safe to give them the foulbrood honey after melting the combs, or would I have to boil it?

A. You must boil it. If you boil it without any water, the outer part will burn while the center is not heated enough to make it safe. So add water, perhaps half as much water as honey, slowly heating at first until all is thoroughly melted, and then bring it to a boil and keep it there for at least fifteen minutes.

Q. Is the wax worth rendering out of the combs of a foulbroody colony, or would it still contain the microbes?

A. The wax is considered all right.

Q. More than once in convention reports, I have read where it was directly stated or intimated that bees do not have foulbrood in trees, buildings, etc., and now A. W. Smyth, in an extract from Irish Bee Journal, says: "No one has found foulbrood in bees * * * in any home not purposely made for them." I should like to know on what this common belief is founded. If this is the rule, I know of at least one exception, as I took a colony of bees from a house, which colony had European foulbrood and I cannot see any reason why such a home for bees should be exempt from the disease.

A. I do not think that the opinion prevails on this side of the water that bees never have foulbrood "in any home not purposely made for them." Indeed it has been urged that one reason why it was so difficult to get rid of foulbrood was because of diseased wild colonies. Why should not a wild colony be exposed to precisely the same dangers as one in a Langstroth hive? Your one case is enough to prove that bees may have foulbrood in a home not specially prepared for them.

Foulbrood, European.—Q. What is the color of European foulbrood?

A. The unsealed larva, instead of being pearly white, as in a state of health, is of a distinctly yellow tinge, becoming darker as it dries, until very dark brown or black.

Q. Tell us how to destroy European foulbrood without destroying a lot of nice, straight combs.

A. When I discovered European foulbrood in my apiary, I melted up hundreds of beautiful worker-combs. If I had it to do over again I would try to save them. I have been blamed for encouraging anything of the kind, because in the hands of careless beekeepers there is danger that the disease may be spread through the combs that are saved. But you'll promise to be very careful, won't you, if I tell you how I would do—how I have done? The first thing is to have the colony strong. Foulbrood is not a great strengthener of colonies, and if it has proceeded to any great extent you will need to strengthen the colony by giving brood or young bees, or both, from healthy colonies, or by uniting diseased colonies. But, remember, the colony must be strong. The Alexander treatment requires the removal of the queen, and then 20 days later the giving of a ripe queen-cell or a virgin just hatched of best Italian stock. The bees do the rest. I think I have had just as good success without leaving the colony so long without a laying queen. So instead of waiting 20 days, give the colony a cell or a virgin queen just as soon as it will accept it after the removal of the queen.

Sometimes you may find only a single bad cell, or perhaps 8 or 10. In that case it may not be necessary to do anything. A week or two later you may find that the bees have cleaned out all bad brood and left nothing but healthy brood in the hive. But you may find the case worse than it was, although not yet a very bad case. If the queen is vigorous, and the colony appears prosperous, cage the queen and leave her in the hive. After a certain period let the queen out of the cage, and if your bees do as mine have done the disease will have disappeared in most cases. I say after a certain period. I think a week is long enough, but perhaps ten days is better. You notice that I also say, "in most cases." Because in more cases than I like the disease has reappeared. But so it did in some cases when I brushed the bees upon foundation and melted the combs.

Q. Colonies have one, two or three cells of European foulbrood, say first of June. If I kill the queen the last half of clover flow and let these bees rear their own queen, will this cure

European foulbrood? If so, state time to do it. Clover flow from June 20 to July 20.

A. A cure would be likely to follow. Better not wait until the last half of the flow, as the case would be getting worse all the time, but act at the beginning of the flow. But if only two or three diseased cells are present, and the queen is good, all you need to do is to cage her in the hive for ten days.

Q. If caging a queen for a certain length of time, in case of European foulbrood, stops the disease, should the disease not come to an end in fall, as all brood-rearing stops entirely for several months?

If an apiary has foulbrood one season, will it be free from it next year? There are no young diseased larvæ from which the nurse bees can suck the juice and feed it to healthy ones the next spring.

A. The shortest answer to your question would be to say I don't know. And that's the truth. I don't know why caging a queen should stop the disease. If caging a queen stops the disease, I don't know why the winter's rest from brood-rearing does not stop it. But here is the important fact that I do know. I know that in a large number of cases cessation of brood-rearing for a week or so has stopped the disease. Note that I don't say in all cases, but in the large majority of cases. I don't know that in the great majority of cases the disease is conveyed from one cell to another by the nurse-bees sucking the juices of recently-diseased larvæ, but it is a pretty satisfactory theory until a better theory is advanced.

I think, however, that no one has advanced the theory that the disease is in all cases conveyed by means of larvæ that have been dead only a short time. It may in some cases be conveyed through spores in dried-up scales of larvæ that have been dead a long time. But I suppose these last cases are exceptional. Now, although I don't know all about it, if you will allow me to theorize, I'll tell you what I think is possible in the case you mention. In early spring or winter, when the brood-rearing begins, there are no diseased larvæ present. But there are dried scales containing spores. One would expect that the disease would begin rather slowly from these. And observation confirms that supposition. In a colony which has not been badly diseased in the previous year, the first examination in the following spring shows very little disease—possibly none. Subsequent examinations will show it on the increase, although if I am not mistaken there are some cases in which a colony will remain healthy which has been slightly diseased the previous year. If a colony has been very badly diseased this year, next year you may look for it at the very start with plenty of diseased larvæ, probably because of the millions of spores that are present.

Q. In treating colonies with European foulbrood by dequeening or caging the queen all agree the first thing to do is to make the colony strong. I find that ideas differ on this matter of strong colonies. What is the minimum strength with which you could expect success?

A. You have struck a new question, yet now that it is asked the wonder is that it was never asked before. Without being dogmatic about it, I should say that the colony should be strong enough to have six Langstroth frames well filled with brood—to be more specific about it, each frame being three-fourths filled. I think it also important that there be a good force of young bees, and without this it would not be likely that six frames would be well filled with brood. Old bees that have begun work afield are not the ones that do house-cleaning, and it may well be questioned whether doubling up such bees to any extent would answer the purpose.

Q. Are the germs of European foulbrood transmitted by honey, or, in other words, would a frame of sealed or unsealed honey (with no brood) from an infected colony infect a healthy one?

A. I think it would in some cases. I know that in some cases it does not. I would have little fear of surplus honey from an infected colony. I would not feel quite so safe about a brood-comb, even if it contained no brood. With American foulbrood the case is different. However, in either case, I should prefer both honey and combs that had never been within a mile of a foulbroody hive.

Q. Would combs that have never contained brood be affected in any way, even if they had been drawn out by colonies affected with European foulbrood? Would it be safe to use any of those combs? Now I have 200 self-spacing frames all drawn-out combs. They have been exposed to the diseased colonies, but not used for brood-rearing. The diseased colonies had stored honey in them, and I extracted it. Would it be all right to use them, or would it be better to make wax out of them? Everybody's bees are affected around here, as one of the beekeepers left his hives out to be cleaned up where bees had died.

A. I have used such combs without bad results. Whether it would always work so well I cannot say. If I had never had the disease I should not want to use them. But in your case, with the disease all around you, and having already been in your apiary I should not hesitate to use them. The likelihood is that it will be some time before you are entirely rid of European foulbrood, but it will gradually become less troublesome, and will not hinder you from getting crops of honey.

Q. You state you will never melt up any more combs on account of European foulbrood. What would you do with combs partly filled with honey, or empty, that were left by a colony that had died with the disease?

A. Candidly, I must confess I don't know. As you state the case, I can imagine a colony so thoroughly rotten with the disease that it dies outright, leaving combs containing some honey, but most of the cells filled with diseased and dead brood. If I had such a case I should feel a good deal like burning up the whole thing. I'm pretty certain I should if it were the only diseased colony in the apiary. If the disease were spread throughout the apiary, I think I would let such bad combs dry until the dead larvæ were dry. Then, if there was honey in some of the combs that I thought fit for table use, I might extract it. Whether the combs were extracted or not, I might give them in an upper story to some colony having the disease but not wholly affected. In fact, this latter is just what I did, piling the diseased combs four or five stories high—only the combs were not so badly diseased as in the supposed case.

Even while saying that, with a single case in the apiary so bad as imagined, I should burn up the whole thing, I will stand by my assertion that I will never melt up any combs on account of European foulbrood, because I am very sure I'll never allow a case to get so bad as supposed.

Foundation (See Comb-Foundation.)

Frames.—Q. Is there any difference in the size of the Hoffman and Langstroth frames? If so, what are the outside dimensions of each?

A. Both the same size—17⅝x9⅛.

Q. Are the self-spacing Hoffman brood-frames the best?

A. If the bee-glue is not troublesome where you are, you will find them excellent. If glue is plenty, they are bad.

Q. Do the metal-spaced frames give ample room for bees to pass between frames?

A. Yes, they take up almost no room.

Q. Which frame do you think is the better, the Hoffman or loose-top, staple-spaced frame, and which is the easier to. handle? What frame do you use, also what size section or extracting-frame?

A. Preferences differ. Some like the Hoffman, and others would not have it around, because the bees glue the frames together, making them harder to handle than the other kinds of frames. With the metal spacers latterly used on the Hoffman, it is not so objectionable.

I use the Miller frame, which is a plain Langstroth frame with common galvanized shingle nails for side-spacers and small staples for end-spacers. I use the same for an extracting-frame, although if I were going extensively into extracting I would likely have a shallower frame. I use the section most generally in use, 2-bee-way., 4¼x4¼x1⅞.

Q. Using plain frames, how do you manage to keep them from swinging and killing the bees when hauling over rough roads?

A. In the same sense you seem to mean, I don't use plain frames. Nothing can be plainer than the Miller frame, except that there are common nails, as I have often explained, used as side-spacers, and staples as end-spacers. Nothing is needed at any time to prepare the bees for hauling, except to close the entrance with wire-cloth.

Q. Would there be trouble with frames made short enough so that there would be a half-inch beespace between the end-bars and the inside of the hive? I have trouble with the standard frame on account of smashing bees. Would the bees fill the space between the end-bars and hive-ends with comb? I use the staple-spaced frame.

A. You would be badly troubled with combs built in such a large space; at least in some cases. Possibly you might like W. L. Coggshall's plan. Drive staples into end-bars at the lower end, so the end-bars cannot crowd against the end-wall of the hive.

Q. In an answer to "Virginia," you tell him to use the wedges that come with the frames. I make my frames. Please explain how to make or get them, and how to use them.

A. A saw-kerf is made in the under side of the top-bar, into which the edge of the foundation goes. Then close beside this is another saw-kerf made by a finer saw, and into this narrower kerf the wedge is crowded. The wedge is a thin strip of wood as long as the under side of the top-bar, one side being chamfered down to an edge, so as to enter the kerf. If you make your own frames it will perhaps be easier for you to have no saw-kerf in the top-bar, but merely let the foundation come up to the top-bar on the under side, and cement it there with melted wax.

Q. How close can frames be together where there are no foundation sheets used? Can they be 1⅛ inches apart? I have them 1⅜ inches, from center to center, and the bees build more combs in a hive than there are frames.

A. You cannot have combs built true without having at least

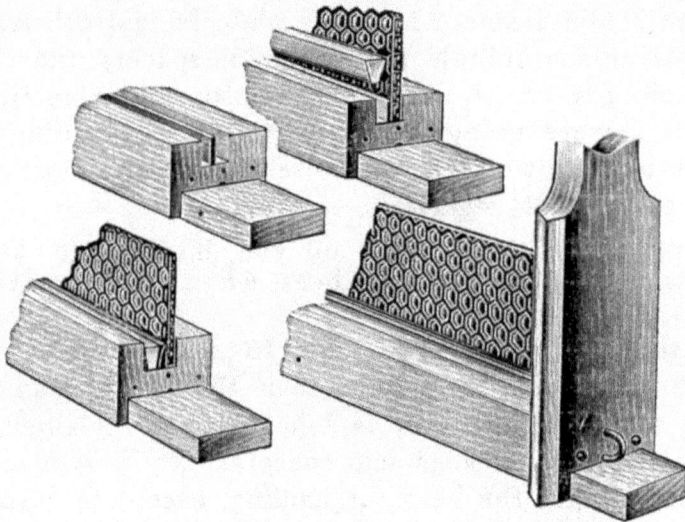

FIG. 18. Section of a grooved frame, showing method of fastening foundation in the top-bar with a wedge.

starters, and full sheets are best, and 1⅜ is close enough. If you try 1⅛ you will find the bees will do still worse than with 1⅜.

The Dadants use frames spaced 1½ from center to center.

Q. Would you advise wiring or putting splints in shallow ex-tracting-frames (5⅜ inches deep), or would they be as well with-out wire or splints?

A. You can get along without any sort of support for the foundation by being more careful in handling the frames and taking a little more time with the extractor, especially while the combs are new. The time of putting in the supports must be figured against the extra time of manipulation without supports.

Q. I suggest, instead of the 10-frame hive being made wider, that the frames be made one-sixteenth of an inch narrower, which would leave five-eighths of an inch more extra room than there is now. So as not to decrease greatly the space between the top-bars it would probably be good to have the top-bars at least one-thirty-second of an inch narrower than they now are.

I believe frames one and five-sixteenths inches wide would be plenty wide enough.

A. I don't believe you would like the plan. If you had loose-hanging frames it might do, and in that case there would be no need to make frames any narrower. But now fixed-distance frames are mostly in use, and 1⅜ from center to center is as little as ought to be allowed. Indeed some prefer 1½. You say, "I believe frames 1 5-16 inches wide would be wide enough." I should say so! I think no one has them wider than 1⅛. Evidently you mean the space from center to center, and, as already said, I don't think you would be satisfied with less than 1⅜. If you should try it, better try it on a very small scale.

Q. What size of extracting-frames are better, the shallow frames or the full depth?

A. The shallow frames are the better, probably in every respect except that they cannot be used interchangeably with brood-combs. Shallow extracting-frames with side-bar 6 inches deep are liked best by the Dadants.

Q. Which is the better extracting-frame, the seven-eighths-inch top-bar with two grooves, or the half-inch top-bar with one groove for extracted and chunk honey?

A. One will probably work as well as the other.

Q. How many frames would you advise putting in a 10-frame extracting-super in order to get nice, thick combs, using full sheets of foundation? I think it is easier to uncap thick combs. Will not the bees build brace or burr-combs if the extracting-frames are too far apart?

A. Either 9 or 8 frames will work well. No trouble with combs built between in either case. If only 8 frames are used, it will increase the space between combs one-third of an inch, and bees will not start an extra comb in so small a space.

Frame, Miller.—Q. Please explain the Miller frame.

A. The frame is, of course, the regular Langstroth size, 17⅝x 9⅛. Top-bar, bottom-bar and end-bars are uniform in width, 1⅛ inches throughout their whole dimensions. The top-bar is ⅞-inch thick, with the usual saw-kerf to receive the foundation, and close beside this is another kerf to receive the wedge that fastens the foundation. The length of the top-bar is 18⅝ inches and ⅞x 9-16 is rabbetted out of each end to receive the end-bar. The end-bar is 8 9-16x1⅛x⅜. The bottom-bar consists of two pieces, each 17⅝x½x⅜? This allows ⅛ inch between the two parts to receive the foundation, making the bottom-bar 1⅛ inches wide when nailed.

The object of the two parts to the bottom-bar is to allow the foundation to come down between them, thus making a close fit without any pains to cut the foundation exactly. After the comb is built in the frame the bottom-bar is no better for being in two parts—perhaps not so good. Some of my frames have a solid bottom 17⅝x1⅛x¼, with the foundation cut to fit exactly down on the bottom-bar. I like them just as well.

The side-spacing, which holds the frame the proper distance from its next neighbor, is accomplished by means of common wire nails. These nails are 1¼ inches long and rather heavy, about 3-32 inch in thickness, with a head less than ¼ inch across. By means of a wooden gauge which allows them to be driven only to a fixed depth, they are driven in to such a depth that the head remains projecting out a fourth of an inch.

Each frame has four spacing-nails. A nail is driven into each end of the top-bar on opposite sides, the nail being about an inch and a half from the extreme end of the top-bar, and a fourth of an inch from its upper surface. About 2¼ inches from the bottom of the frame a nail is driven into each end-bar, these nails being also on opposite sides. Hold the frame up before you in its natural position, each hand holding one end of the top-bar, and the two nails at the right end will be on the side from you, while the two nails at the left end will be on the side nearest you. The end-spacing is done by means of the usual staple, about ⅜ inch wide.

Q. How is the foundation fastened to the top-bar?

A. I prefer what is the usual way at the present time, as suggested in the foregoing description, the foundation being received in a saw-kerf and wedged there, but it can be fastened in any other way.

Q. I would like to ask your opinion, after reading your book entitled "Fifty Years Among the Bees." On page 83 you give the dimensions of your frames, and further on you mention splints, which I think I would like. How would it do to make the bottom-bar the same thickness as the top-bar, and instead of having two grooves, one for foundation and one for wedge, have only one groove in each bar? Then by having a board nearly the same size as inside of frame, and thick enough to come to bottom of grooves, the foundation, by buckling a trifle, could be made to enter grooves? After boiling the splints in wax, buckle them into place the same as foundation. Then use hot wax along the top and bottom-bar to fasten it in. This would reduce the size of frame, but with the Hoffman frames I find it hard to get the bees to build down to the bottom bar as they should, so lose some space there anyway.

A. Your plan will work all right. But you don't need to have any kerf in the bottom-bar, and then you won't need to have it as thick as the top-bar. Indeed, if you wax the foundation, top and bottom, you will not need kerfs either place. I have some frames without the split bottom-bar, and it works all right. You may say you want the kerf to hold the splint. I never yet put a splint in a kerf, and see no need of it. Of course, the top-bar must be thick, kerf or no kerf.

.Q. What is your opinion of the use of the Miller or "metal-spaced" frame with top-bars seven-eighths inch square for any location, either comb or extracted honey, the idea being that the combs could be trimmed to the proper thickness with the narrow bar, while the knife would not work against the metal or nails, and at the same time the frame might be used for producing comb honey?

A. It might work satisfactorily; but only after trial could one be sure about it. How much the metal spacers would be in the way of an uncapping-knife would depend upon their construction. If there is metal at each end on each side, there would be trouble. As you know, I use common nails as spacers. These are only on one end on each side of the frame, and by starting the knife at the end where the spacers are, there ought to be little danger of striking the knife on the metal. I have seen in foreign bee-papers mention of metal spacers that were removable, being taken off for extracting, and then put on again upon returning to the hive.

Q. How thick should the follower be in order to hold the frames solidly together?

A. Strictly speaking, the frames are never held solidly together. They are crowded closely together against one side, but there is left a loose space at the other side between the dummy and the side of the hive. No possible harm can come from this except that it allows a little movement when hives are hauled over very rough roads, but I have never had any trouble in that way. The dummy is five-sixteenths of an inch thick.

Q. How wide were the top-bars of the unspaced frame formerly in use by you?

A. Seven-eighths of an inch.

Q. Did you find disadvantages in the unspaced frame other than those mentioned in your book?

A. Yes, there was at least one other. As the frames hung entirely free, in time there was a little warping of some of the top-bars. Every slight twist of the top-bar would allow quite a bit of swing out of true at the bottom of the frame, so that it sometimes happened that at the bottom, the end-bars or bottom-bars were glued together, a very unpleasant annoyance.

Freezing of Bees.—Q. Do bees often freeze to death with plenty of stores?

A. No; unless the colony is too weak or a small cluster of bees gets caught in a cold spell away from the main cluster.

Fruit, Bees Injuring.—Q. Do bees injure sound fruit?

A. No, they do not and cannot, since the mandibles of the honeybee are rounding, and cannot pierce the skins of sound fruit. Tests of this were made at the Ottawa Experiment Station in Ontario, Canada. First, strawberries were tried, then raspberries, neither of which were injured.

The fruit was placed inside the hives, also in other places easy of access to the bees. Inside the hive the fruit was exposed in three different positions.

(1) Whole fruit without any treatment.

(2) Whole fruit that had been dipped in honey, in one half the super.

(3) Punctured specimens in the other half the super.

A second test of the same kind was made with peaches, pears, plums and grapes.

"The bees began to work at once both upon the dipped and punctured fruit. The former was cleaned thoroughly of honey during the first night; upon the punctured fruit the bees clustered thickly, sucking the juice through the punctures as long as they could obtain any liquid. At the end of six days, all the fruit was carefully examined. The sound fruit was still uninjured in any way. The dipped fruit was in like condition, quite sound, but every vestige of honey had disappeared. The punctured fruit was badly mutilated and worthless; beneath each puncture was a cavity, and in many instances decay had set in. The experiment was continued during the following week, the undipped fruit being left in the brood-chamber; the dipped fruit was given a new coating of honey and replaced in the super, and a fresh supply of punctured fruit was substituted for that which had been destroyed.

"After the third week the bees that belonged to the two hives, which had been deprived of all their honey, appeared to be very sluggish, and there were many dead bees about the hives; the weather being damp and cool was very much against those colonies. These colonies had lived for the first three weeks on the punctured fruit and on the honey off of the fruit which had been dipped; as there were at that season few plants in flower from which they could gather nectar, these bees had died of starvation, notwithstanding the proximity of the ripe, juicy fruit. The supply of food which they were so urgently in need of was only separated from them by the skin of the fruit, which, however, this evidence proves, they could not puncture, as they did not do so."

Fruit-Bloom.—Q. Is there much honey from fruit-bloom (principally apples)?

A. I am in a region of abundant fruit-bloom. but I never had a pound of surplus from it. It is all used up in rearing brood. If it came in the middle of June I should probably have had tons of honey from it. Yet I wouldn't for many dollars have it in June. The bees reared from fruit-bloom are what gather the surplus later on, and so fruit-bloom is of the highest value. In this region apple is worth all the rest put together, for it lasts two to four weeks, there being that difference between the earliest and latest varieties.

Gentle Bees.—Q. What strain or race of bees do you consider most gentle and easy to handle?

A. The Caucasians are claimed to be gentlest of all, but reports do not all agree. Italians are good.

German Bee-Papers.—Q. Is there a German bee-paper published, either here or in foreign countries?

A. No German bee-paper is published in this country, but a number across the water, among them Schweizerische Bienenzeitung, Praktischer Wegweiser, Leipziger Bienenzeitung, Bienen-Vater, Deutsche Imker aus Boehmen. Names and addresses of German papers can probably be obtained by addressing request to the office of the American Bee Journal.

Giant Bee of India.—Q. Do you think the giant East Indian honeybee will ever be imported to this country?

A. No; and I don't believe it would be of any value if it were brought here.

Glass for Super-Covers.—Q: I have noticed two or three times in the American Bee Journal beekeepers using a sheet of glass for a super cover. I would like to adopt it myself if it would be advisable, but before deciding would like to have your opinion for and against it.

A. Some have reported success in using glass over the brood-chamber, especially in England, while others object to it. I'm not sure what the objection is, but suppose there would be trouble with vapor condensing on the glass and dropping down upon the bees. The advantage is that you can see through the glass, yet there is not so very much to be seen without lifting out anything. You will probably be wise not to try it on a very large scale, at first.

Gloves.—Q. What kind of gloves do you think best for handling bees? Will bees sting through kid gloves?

A. Bees will sting through anything as thin as kid gloves. Buckskin does better, but is not always proof against stings. Rubber gloves are good, but uncomfortable. Pigskin is probably as good as anything, and not expensive. It has a disagreeable smell. especially when new. Oiled cotton gloves are in common use, and do very well.

Glucose.—Q. Is syrup that contains 95 per cent glucose and 5 per cent sorghum good to feed to bees?

A. No; commercial glucose is not fit stuff for man or bee. Don't think of giving bees glucose in any proportion whatever, any time.

Goldenrod.—Q. Does the goldenrod yield honey, that is, does it yield enough for a surplus?

A. Yes; in many sections of the central west and of the east, goldenrod is a surplus producer. In other localities it only helps in the fall flow. Its honey is said to be of a golden color and of a rather strong taste. There are many varieties and some do not yield honey anywhere.

Grading Rules.—Q. By what standard is honey (comb and extracted) graded for the market?

A. The Colorado grading rules, as adopted by the Colorado Honey Producers' Association, come as near being the standard as any. They are as follows:

COMB HONEY

Fancy—Sections to be well filled, combs firmly attached on all sides and evenly capped, except the outside row next to the wood. Honey, comb and cappings white, or slightly off color. Combs not projecting beyond the wood, sections to be well cleaned. No section in this grade to weigh less than 12½ ounces net or 13½ ounces gross. The top of each section in this grade must be stamped, "Net weight not less than 12½ ounces."

The front sections in each case must be of uniform color and finish, and shall be a true representation of the contents of the case.

No. 1.—Sections to be well filled, combs firmly attached, not projecting beyond the wood, and entirely capped, except the outside row, next to the wood. Honey, comb and cappings from white to light amber in color. Sections to be cleaned. No section in this grade to weigh less than 11 ounces net or 12 ounces gross. The top of each section in this grade must be stamped, "Net weight not less than 11 ounces." The front sections in each case must be of uniform color and finish, and shall be a true representation of the contents of the case.

No. 2.—This grade is composed of sections that are entirely capped, except row next to the wood, weighing not less than 10 ounces net or 11 ounces gross. Also of such sections that weigh 11 ounces net or 12 ounces gross, or more, and have not more than 50 uncapped cells altogether, which must be filled with honey. Honey, comb and cappings from white to amber in color. Sections to be well cleaned. The top of each section in this grade must be stamped, "Net weight not less than 10 ounces." The front sections in each case must be of uniform color and finish, and shall be a true representation of the contents of the case.

Honey that is not permitted in shipping grades is as follows:
Honey packed in second-hand cases.
Honey in badly stained or mildewed sections.
Honey showing signs of granulation.
Leaking, injured or patched up sections.
Sections containing honeydew.
Sections with more than 50 uncapped cells or a less number of empty cells.
Sections weighing less than the minimum weight.
All such honey should be disposed of in the home market.

EXTRACTED HONEY

Extracted honey is classed as white, light amber and amber; the letters "W.," "L. A.," "A." should be used in designating color, and these letters should be stamped on top of each can. Extracted honey for shipping must be packed in new, substantial cases of proper size.

Extracted honey should be thoroughly ripened, weighing not less than 12 pounds per gallon. It must be well strained and packed in new cans; 60 pounds shall be packed in each 5-gallon can, and the top of each 5-gallon can shall be stamped or labeled, "Net weight not less than 60 pounds."

Strained honey must be well ripened, weighing not less than 12 pounds per gallon. It must be well strained, and if packed in 5-gallon cans each can shall contain 60 pounds. The top of each 5-gallon can shall be stamped or labeled "Net weight not less than 60 pounds." Bright, clean cans that previously contained honey may be used for strained honey.

Honey not permitted in shipping grades is as follows:

Extracted honey packed in second-hand cans.

Unripe or fermenting honey, weighing less than 12 pounds per gallon.

Honey contaminated by excessive use of smoke.

Honey not properly strained.

Honey contaminated by honeydew.

Granulation of Honey.—Q. What causes the granulation of honey? Is there any way to prevent it? I sold some to a man this fall. He says it was granulated in the combs and he will not buy any more. Does it make any difference when the honey is gathered from different flowers as to its "sugaring"?

A. The granulation of honey is caused, or at least hastened, by cold. Some honey, however, granulates readily without being reduced to a low temperature, since the honey from some plants granulates very readily, while the honey from some other plants scarcely granulates at all. Frequent changes from warm to cold favor granulation more than a steady continuance of cold. Stirring honey hastens granulation. If honey is heated as much as it will stand without injuring the aroma or flavor, say somewhere below 160 degrees, and sealed up while hot, it will continue liquid.

Your inquiry, however, is more particularly about comb honey. While honey in the comb is slower about granulation than extracted honey, we are more helpless about preventing granulation or reducing it to a liquid state after it is once granulated. To be sure some have reported melting comb honey—or bringing it again to a liquid state—without injuring the comb, yet it must be a rather ticklish job. I think that honey left a considerable time on the hive is less inclined to granulate than that which is removed just as soon as it is sealed, but here you meet the trouble that leaving it on the hive too long darkens the comb. Perhaps the best you can do is to leave your sections on as long as you can without having the combs darkened, and then keep them in as warm a place as you can until sold.

Q. Under what conditions can extracted honey be most quickly granulated, or candied, so that it can be sold in paper packages? I am not engaged in beekeeping, and haven't much literature on the subject.

A. In Europe, where there is more desire to have honey granulate than here, they stir the honey occasionally. Mixing a little granulated honey with the liquid also helps. There is a great difference in the kinds of honey. Some honey begins to granulate as soon as extracted, while other honey may remain liquid a year or more.

Q. Can combs containing granulated honey be fed to the bees in the spring? If not, what can I do with them?

A. You can give them to the bees, but unless some precaution is taken they will throw out the granules and waste them. Sprinkle them with water, then give them to the bees, and as often as they lick them up dry, sprinkle them again.

Grapes.—Q. In central California the grapes are sour (not much sugar) and my bees have gathered some of this juice, consequently the honey has a somewhat sour taste. Is this good winter feed for the bees or for consumption? (California.)

A. My guess is that it will not be good for winter stores. It will be all right for consumption if the taste is not objectionable, and of that you can judge better than I. The same may be said of all fruit juices.

Guards (See Entrance Guards.)

Handling Bees.—Q. How warm should it be by the thermometer when it is safe to handle bees in ordinary manipulation?

A. About 70 degrees. Instead of going by the thermometer, it may be better to say, don't handle bees any time when they are not flying freely. But if you merely lift out a frame and quickly return it, as when you want to know in the spring whether brood is present, then it may be safe at 55 degrees or less.

Q. I have a colony of bees that I have left outside with a box cover packed with leaves. They have nothing over the brood-frames, but are wintering finely. Does it hurt the bees much to open the hive in cold weather?

A. Sometimes it does a great deal of harm, even to the death of the colony, to open the hive and disturb the bees when it is too cold for them to fly. When it is warm enough for them to fly, it may do little or no harm; but when very cold, better not disturb them unless there is danger of starvation.

Hanging Out of Bees.—Q. My bees have been hanging from the top of the hive to the ground. They fly around the hive and then cluster. Only a few seem to work. They have been doing this for two weeks. Are they getting ready to swarm?

A. I don't know enough about the conditions to answer. If no nectar is to be had, that may be a sufficient reason for their idleness. If there is a good flow of nectar, hanging out might be a sign they are getting ready to swarm, and yet they would hardly keep that up for two weeks. So, on the whole, it looks more as if there is nothing for them to do, yet that may not be the case at all. **Give them more room, more ventilation and more shade.**

FIG. 19. A colony "hanging out" all over its hive; caused by lack of room, lack of shade, and insufficient ventilation.

Hatching.—Q. My attention has been called to the word "hatch." Do bees hatch more than once? Would it not be better to have the bees **hatch** once and **emerge** to come into existence?

A. You are quite right; it would be better to say that the larva "hatches" from the egg, and the young bee "emerges" from the cell. Indeed, you will find that quite often the word "emerges" is used in that way, although generally it is said that the young bee hatches out of the cell.

Q. How long is it from an egg to a bee? I mean how long after the egg is laid till it is a full-grown bee?

A. For a queen, 15 or 16 days; for a worker, 21 days; for a drone, 24 days.

Heartsease.—Q. My bees have done well in this part of Southern Kansas this season. There is no trouble in wintering here, as they have a flight nearly every week. I expect to move to south-central Iowa this fall. Will my bees winter successfully there on heartsease honey, or would it be better to extract the honey from the brood-nest and feed sugar syrup? In 1905, some of my bees died of dysentery wintering on heartsease honey and not being able to take a flight for about six weeks on account of the severe weather.

A. It is possible that heartsease honey was not to blame for the trouble of 1905. Surely thousands of colonies have wintered on it, and it has not had the name of being bad for winter food. My bees wintered well last winter, and I think a good share of their honey was heartsease.

Hive-Stand.—Q. Which is better, a hive-stand a couple of feet high, or one a few inches high, with the entrance-board slanting, so that in case the clipped queen went out to swarm she could crawl back in the hive again and thus not be lost?

A. For you it may be better to have the hive quite low. Where certain kinds of ants are bad (generally in the south), it is well to have the hive on legs so that by means of dishes of oil or water the ants may be prevented from getting into hives.

Hives.—Q. I would like to know if there is a book on making hives?

A. I know of no such book.

Q. I am not a young man in years, but am young in the knowledge of bees. I keep bees only for the honey I can get. What use is there for me to use patent hives when I know nothing about them? Why is not my old-fashioned gum with a good, big, plain box-cap just as good for getting the same amount of honey in a season as the patent hives?

A. Let me say, first, that most of the hives in use now by practical beekeepers have no patent on them, the patent on the Langstroth movable frame having expired many years ago. So your question probably is: What advantage is there for you in movable-frame hives over common box-hives? Perhaps there is no advantage. It depends upon circumstances. The movable-frame hive is no better for the bees than a box-hive; in general not so good. It has really only one advantage over a box-hive, but sometimes a single advantage counts for much. A man with his head on has the single advantage over the one with his head cut off that he still has his head on; but that is a considerable advantage. The one advantage that the movable-frame hive has over the box-hive is that the frames can be taken out and put back again. But that advantage is of no value to those beekeepers who never lift out the frames from one year's end to the other. If I had no notion of ever lifting out a frame I would prefer box-hives.

Possibly you may want to know what advantage there is in being able to lift out frames. For one thing, you can tell by lifting out the frames whether a colony is queenless or not, and if it is queenless you can remedy it. With a box-hive it is practically impossible either to detect or to cure queenlessness. That one difference between the two kinds of hive is enough to decide in favor of the movable-frame kind, provided one intends to take advantage of the movable feature. It would be a pretty long story to tell all the things that can be done with a movable-frame hive that cannot be done with a box-hive, among which are examination for disease and treating for the same, introducing queens, strengthening weak colonies by giving frames of hatching-brood, etc.

Q. Please give me some advice on what kind of hives to use.

A. Opinions differ as to what is the best hive. Some are partial to this or that particular hive which the majority of beekeepers would hardly take as a gift. The greater number, however, perhaps nine out of every ten, would tell you to take the 10-frame dovetailed hive. You can hardly go amiss on that. But please remember that the hive does not make very much difference in the work of the bees. A good colony of bees will store just as much honey in an old-fashioned straw hive as it will in the most up-to-date hive. But it makes a big difference to the beekeeper whether the hive is such that he can easily get at the honey and perform the various manipulations that he may think necessary.

Q. What are the dimensions of an 8-frame Langstroth hive and super? Also the frames?

A. Some of the dimensions of the 8-frame hive have varied from time to time, but I'll give you what I think will generally be found today:

Length, inside measure, 18¼ inches; width, 12⅛; depth, 9⅜;

but, as the dryest lumber you are likely to get will shrink some-what, it is better to make the depth 9½. The super has the same length and width as the hive. Its depth depends upon what it contains. If it is an extracting-super, it will be the same as the hive-body, provided the frames are to be the same as those in the brood-chamber. In any case, the depth of the extracting-super will be one-fourth inch more than the depth of the frame to be used in it, allowance to be made for shrinkage, if there is to be any shrinkage. The depth of the section-super must be such that there shall be one-fourth inch space left at the top of the super.

The frame is 17⅝ by 9⅛, outside measure. Width of top-bar varies from 1⅛ down to ¾; and the same may be said of end-bars and bottom-bar. Some have the same width as the top-bar, and some have them narrower. In any case, the frames are generally spaced so that the distance from center to center shall be 1⅜, although some prefer 1½. With the spacing 1⅜, there is plenty of room for a thin dummy or follower beside the frames.

Q. What are the exact measurements of a 10-frame hive, in-side measure?

A. Unfortunately, there are no "exact measurements" that all makers have always used in making hives to take 10 frames of Langstroth size. The depth of the frames being 9⅛ inches, if ¼ inch be added to that to make a beespace, we would have 9⅜ for the depth of the hive. But a very little shrinkage would make bad work, and to make sure against that, the hive is made 9½ inches deep. The length of the frame is 17⅝, and if ¼ inch be added at each end we would have 18⅛ for the length of the hive. But that makes very close work, and bees are not much inclined to build at the ends of the hive, so the length is not less than 19¼. For an 8-frame hive I think there is general agreement on 12⅛ for the width. That allows 11 inches for the 8 frames spaced 1⅜, and 1⅛ inches for a dummy ⅜ thick, with a space each side of it. If we add twice 1⅜, or 2¾ inches, for two additional frames, we would have 14⅞ for the width of a 10-frame hive. But for some reason that never seemed satisfactory to me, the dummy is generally omitted in 10-frame hives, and they are made 14¼ inches wide. So I think we may say, as nearly as we can come to standard, that the inside measurements of the 10-frame hive are, 18¼x14¼x9½. As a side remark, I may say that I think some of the hives are not more than 9⅜ deep, although I think they were 9½ when new.

Q. Do you think the bees will gather more honey in a 10-frame hive than they will in an 8-frame?

A. Not necessarily. Of course, a stronger colony ought to get more honey, but just as strong a colony can be in an 8-frame hive as in a 10-frame, for two stories of the 8-frame can be used if need be. Of course this would be a 16-frame hive.

Q. I would like to ask a few questions concerning that large hive.

(a) When you take that second hive off, don't you have trouble with brood, or do you use an excluder?

(b) Do your bees go to work in the supers as readily as when only one hive-body was used?

A. (a) I use 8-frame hives which can hardly be called "large hives," so I suppose you refer to my using two stories as brood-chambers, making practically a 16-frame hive. I put on a second brood-story whenever the first becomes crowded, unless I take away some of the brood to use elsewhere. I reduce to one story at the time of putting on supers for surplus. There is so little trouble with brood in sections that I don't think it worth while. to use excluders. But if I didn't fill the sections full of foundation, I should have to use excluders.

(b) Yes, perhaps more readily.

Q. What kind of a beehive do you prefer, without porch or with porch, and why?

A. The Langstroth hive was at first made with a portico. Latterly very few have the portico, perhaps chiefly because it furnishes such a nice refuge for spiders, causing the death of too many bees.

Q. Is a hive supposed to sit level?

A. It should slant a little to the front, the front end being an inch or two lower than the back end. It should be level from side to side.

Q. I have seven colonies of bees, four in 8-frame hives and three in 10-frame hives. At 1:30 today the bees of all the smaller hives were flying, while the others were not. I examined them, and found one of the colonies dead, though there was about 30 pounds of nice honey left. In the dead colony there was a double handful of bees and lots of drones.

I cannot account for this. It looks as if 8-frame hives were better for wintering in northern Iowa than the 10-frame. All the colonies had plenty of honey. (December 28, Iowa.)

A. The number of colonies is rather small to deduce a general rule; but even if you had a larger number it is not a dead open-and-shut affair that the smaller hives are better winterers. As to that dead colony in the 10-frame hive, it's about certain that the size of the hive cuts no figure. They had no normal laying queen and had not had one for weeks, for the dead bees were few, and part of them were drones. We have left, then, the four 8-frame hives, and the two 10-frame, and you are evidently of the opinion that the bees in the latter two were in too poor condition to fly, while the bees in the smaller hives flew well. Well, as there were only two of them, it might just happen that those two were poorer than the others. But did it never occur to you that it might be that those two colonies did not fly **because they were in too good condition to fly?** That would be my guess. December 28 the bees had not been confined very long, and these two colonies were doing so well that they did not yet feel the need of a flight. At any rate, wait until spring, and then you can tell with more certainty which has done the better.

Q. Will bees go into old hives as well as into new ones when they have become damp inside several times, but have never been used before?

A. Yes, if the hives are sweet and clean.

Q. For a beginner, which would you recommend, the 8-frame Jumbo brood-chamber, or the 10-frame Langstroth?

A. The 10-frame Langstroth.

Q. The cuts showing how to nail dovetailed hives, nail only every other dovetail. Do you think that is the best way, or should every one be nailed?

A. At top and bottom nail at least two consecutive dovetails; it's not so important about the central ones. I have had pretty good success by driving a nail vertically at top, and one at bottom.

Q. Which hive do you recommend for a beginner, the Tri-State, Dovetailed, or Leahy telescope? Also which number of frame, 8 or 10? (I am located in northeast Missouri.)

A. These all have the regular Langstroth frame, $17\frac{5}{8} \times 9\frac{1}{8}$, the size to be recommended, and aside from this the particular form of hive does not matter greatly. The dovetailed has the advantage that it is the one most generally in use. As to the number of frames, the 10-frame is decidely better.

Q. Is it not easier to cut out queen-cells in the Danzenbaker hive than in the Langstroth? I am pretty badly smitten on the

Danzenbaker hive, but I see you do not like it as well as the 8-frame Langstroth.

A. I know of no reason why it should be easier. If you take into account taking out and putting back frames, it is harder.

Q. I have been looking through the American Bee Journal for dimensions of the Dadant hive and frame about which I wrote

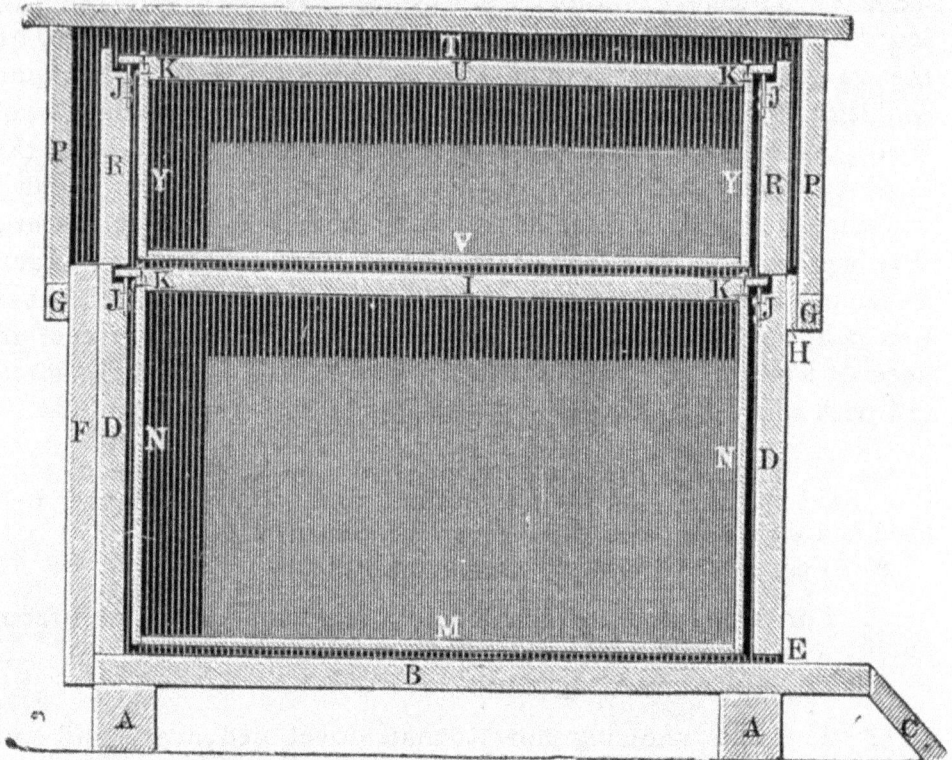

FIG. 20. Cross-section of the Dadant hive as taken from "Langstroth Revised."

Mr. Dadant some months ago, but cannot find them. Will you kindly give them in the replies to queries? The expense of getting a sample hive here is too great.

A. The dimensions of the Dadant hive are not given in any previous number of the American Bee Journal; they are to be found only in the Langstroth-Dadant book and in Bertrand's "Conduite du Rucher" (Conduct of the Apiary), which has been published in eight different languages. The dimensions of the frame are about the same as those of the original Quinby movable frame. The hive is especially adapted to the production of extracted honey, and that is why it is very much more widely used in the countries where modern beekeepers can secure almost as much for extracted per pound as for section honey.

Q. Kindly give the manipulations of divisible brood-chamber hives.

A. Perhaps no two who use divisible hives manage them exactly alike. In a general way, I may say that advantage of divisible hives is taken by reducing to a single story at time of giving supers, although some make the first and second stories exchange places. This last, you will see, throws the honey that was above the brood-nest right into the middle, and the bees are supposed to get busy carrying it up into the supers for the sake of getting brood in its place.

Q. I have decided that a divisible hive consisting of shallow frames and supers, one, two or three, according to the strength of the queen, is about what I want. Is it a practical combination? It looks to me like this hive will be extremely easy of manipulation and that the job of queen and queen-cell finding will be minimized. I wish to winter out of doors, and think I can make a warm hive of the shallow frames and supers by contracting the brood-nest horizontally with a tight division-board on each side and packing between them and the outside; the ends being closed.

A. I doubt the advisability of your trying shallow or divisible-chamber hives. To be sure, some good beekeepers use them, but the majority of beekeepers prefer a frame not less than the Langstroth, and some like a still larger frame. If you do decide to use some of the divisible hives, try only a very few at first, until you decide whether they are suited to your use.

Q. Which is better in a double-walled hive, a dead-air space, or planer shavings packing?

A. It is generally considered better to have packing in the space. Theoretically, air might be thought a better non-conductor than shavings, and so it is if the air would remain still; but the trouble is that it will not remain still, but when a part of it becomes warm, at the warmest part it travels to a cooler part to give up its heat there. The packing stops it from traveling so much.

Q. Is a "chaff" hive entirely practical? If not, what are the objections to it? I have no cave and do not like to contemplate the work incident to packing 50 or more hives with paper or other material.

A. Chaff hives have been successfully used to quite a large extent, although perhaps not so much as formerly. One objection is their weight and unwieldiness; another that when the sun shines on a hive in winter it takes too long for the heat to penetrate the thick walls. They are much used in northern states.

Q. (a) Do the double-walled hives produce more honey than the single-walled?

(b) Are bees wintered out-of-doors better in the former hive than in the latter?

A. (a) No; and in general it may be said that differences in hives are more for the convenience of the beekeeper than for the bees. Looking at it in another way, however, if two hives stand side by side, one with double walls and the other with very thin walls entirely unprotected out-doors in a very cold climate, it might be said that more honey would be produced in one than the other, because the bees might nearly all die in one and not in the other in winter, and that would make a difference in the amount stored the next summer.

(b) Not if the single-walled hive is well packed.

Q. I use the Acme hive and Wisconsin style. Which is better, in your judgment?

A. I like the Wisconsin the better of the two, because it has the regular Langstroth frame. The dovetailed is still better, because the portico of the Wisconsin makes a good shelter for spiders.

Q. Would a 12-frame hive be all right to use here in northern United States? Would the bees swarm as much as they do in 8-frame hives, or would it prevent swarming?

A. Some use 12-frame hives with great satisfaction. Although they will not prevent swarming entirely, there will be much less swarming than with 8-frame hives, and with them you should get as much honey.

Q. Kindly refer me to any bee-papers or other sources of information about the Long-Idea hive.

A. I don't know just where to refer you, although years ago there was quite a little scattered through the bee-papers about the Long-Idea hive. Although used somewhat largely in Europe, it is used very little in this country. O. O. Poppelton is its chief apostle, a very able beekeeper of Florida, who likes it much. All there is of it is to make the one story large enough to contain all the frames you want, so as to use no second story. Some use a queen-excluder so as to separate the hive into two compartments, one for brood and the other for honey. I'm not sure about it, but I rather think Mr. Poppleton does not use this excluder.

Holy Land Bees.—Q. Are the Holy Land bees a different kind from the others, or are they a substitute under a different name? I would like a description of them—color, etc.

A. The Holy Land is the same as the Palestine, and comes from Palestine. They are distinguished as being very prolific, and for starting and maturing a great number of queen-cells; but for some reason they seem not to be in general favor.

FIG. 21.   Blossom and stalk of the bitterweed of the South.   It is the cause of most bitter honey.

Honey, Bitter.—Q. In this vicinity, 30 miles north of Chattanooga, all of the honey stored before May 20, this year, was decidedly bitter. Some say it was peach bloom, some black gum, some dogwood. Do any of these cause bitter honey?

I have been inclined to think the bitter honey came from the bitterweed, or yellow fennel, which was stored in the brood-chamber last September, as there was lots of it in this section last fall. Some of my colonies stored as much as 20 or 30 pounds apiece in supers. It was as bitter as quinine. I fed it to weaker colonies. Could this have been removed from the brood-chamber and carried into the super, as they wanted to make room for the brood, and mixed with other honey?

A. I'm rather glad to live where I have no chance for practical knowledge of such objectionable honey.

Generally, the honey in the brood-chamber is used up for brood, but if the queen were crowded for room the bees might carry honey from the brood-chamber into the super to make room for her.

Helenium tenuifolium, also called "bitterweed" and "sneezeweed," yields bitter honey, but it is not the same as fennel (Anthemis cotula), which is a chamomile and yields no honey.

Honey, Bottled.—Q. What do you think of the plan of bottling honey and making it an expensive luxury so that the consumer can just taste of it occasionally? Would there not be more of the spirit of "loving our neighbors as ourselves" to cut out the middle system of bottling and sell it to him at a figure so that he can make it an article of everyday diet? In the long run, would there not be more dollars and cents for the beekeeper?

A. The way to do is to sell honey in as large and inexpensive containers as possible so as to make as little expense as possible for each pound sold. That ought to give the consumer the most for his money and the producer the most money for his honey. Unfortunately, however, we are often controlled by conditions and circumstances. A large part of the consuming public is in the habit of buying in small quantities. A Chicago retail grocer who should keep honey only in 20 to 60-pound packages would probably sell very little honey; whereas, plenty of customers will buy a pound at a time, even if they must pay for a bottle of no value to them. What better can he do than to keep the small packages?

Honey, Color of.—Q. I would like to know the cause of dark honey.

A. The color of honey depends upon the source from which the bees obtain the nectar. From buckwheat they get honey that is very dark, from fireweed that which is very light, and varying grades from other plants. Sometimes there is a difference in the shade of the same kind of honey obtained in different regions or on different soils. Some alfalfa honey is a shade darker than the lightest to be found elsewhere.

Q. The bees are all storing dark honey, and it has a strong flavor. No one seems to know what causes it, as we have lots of white clover, and also lots of rain.

A. The trouble may be honeydew, and there is no remedy, unless it be to take off all surplus arrangements at the beginning and end of the honeydew flow. Indeed, it is the same if the dark honey comes from any other source.

Honey, Foaming.—Q. In a covered tank I have honey harvested a few months ago. Whenever I dip it up with a spoon and fill a glass it becomes all foamy and runs out of the glass. Does it show that the honey is not ripe? I don't remember if it was all sealed.

A. The probability is that the honey was very unripe.

Honey, Freezing.—Q. It is the custom here in Russia to keep and sell honey in wooden tubs without any covers. Usually it granulates in October or November. It is kept all winter in buildings without stoves, where the temperature is under freezing point. Does freezing injure honey?

A. Freezing does not in any way injure granulated honey. It hastens the granulation of liquid honey, and may crack the combs of comb honey.

Honey, Harvesting.—Q. What month can honey be harvested? (New York.)

A. Comb honey is generally ready to be harvested whenever it is fully sealed over. That probably means in your locality that most of it will be taken off in July and August and still later if there is a late flow. The same rule applies to extracted honey, only some of the best beekeepers prefer to leave all on the hives until the close of the season.

Q. A day or two ago I removed a super of honey, either basswood or sumac, which was entirely sealed over. Upon tasting the honey I found it left a raw taste in my mouth. I suspect it was green, and gave it to the bees again. How can I tell when it is ripened?

A. Generally honey is ripe when it is sealed, and it may be that the objectionable taste comes from some peculiar plant. If that be the case, the bad taste may or may not disappear. Indeed, basswood itself has the reputation of a raw taste until it has attained a certain age, and that taste may disappear, even if the honey be off the hive. I know of no way you can tell when it is ripe except by the taste and the consistency.

Q. (a) When is the best time to take honey from the bees, at noon, in the morning, or in the evening?

(b) How can I kill bees and save the honey? I have two little swarms that are not worth keeping.

(c) How can I get the bees separated from the honey after it is taken off?

A. (a) That depends somewhat on circumstances. Generally

beekeepers take comb honey from time to time as fast as each super is finished and sealed, or nearly so. In that case most of it is taken during the season that bees are busy at work, and it is better to operate while most of the gatherers are abroad in the field, and not so early in the morning or late in the evening. If, however, bee-escapes are used, they are put on toward the after part of the day, and the honey taken before the middle of the next day.

Much the same thing may be said about taking extracted honey, although some of our best practitioners do not take their extracted until the close of the season for each kind of honey. Of course, it is also true that the last of the comb honey is taken at the close of the flow. At such times there may be some gain by getting at work pretty early in the day, before robbers are much on the wing. Those who are in the business extensively do not pay much attention to the time of day, but work away any time of day, or the whole day, just as suits their convenience.

(b) The usual way to kill bees is with the fumes of burning sulphur. But if each of those colonies is too small to be worth saving you may be able to make one fair colony out of the two. Or, you could add each one to some colony that would be the better to be a little stronger. Nowadays it is not usually considered good practice to kill bees.

(c) There are various ways of getting bees out of surplus honey. Some use the Porter bee-escape. Some drive part of the bees out with smoke then pile up the supers on the ground and set a Miller escape on top of each pile. Some simply brush them off the extracting-combs. For a small quantity you can put the honey in a large box, put a sheet over it, and turn the sheet over from time to time as the bees collect on it.

Q. I have several supers of fine honey all capped over and finished. Would you advise me to take it off and put it in a well-ventilated room, or leave it in the hive? If the latter, how long?

A. Take it off as soon as finished. The honey will be as good or better if left on longer, but the comb will become dark. Keep it in a dry, warm place.

Honey, Keeping.—Q. Can honey from this year be kept till next year without spoiling?

A. Yes, there is no trouble in keeping extracted honey over, and even comb honey may be kept in a dry, warm place.

Q. In reply to the question when to take off supers, you say,

"Take off each super when it is full." Now, will you please tell me how to take care of the honey after taking it off, until I can sell or eat it? If I take the super off and put it, no matter where, the ants get at it.

A. Keep the honey in a warm, dry, airy place. If warm and dry it doesn't matter so much about being airy. A place where salt will keep dry, and where it never freezes is a pretty good place. One way to keep it from ants is to have it closed in something so tight-fitting that ants cannot get to it. That's a hard thing to do, especially with a large quantity. An easier way is to put it on some kind of platform supported on four feet, each foot resting in some old dish or can kept supplied with some kind of oil or water. Perhaps you can kill off the ants. If you can trace them to their nest, you can give them a dose of bisulphide of carbon, or gasoline. You can wring a sponge out of sweetened water and put it where the ants will collect on it, then dip ants and all in boiling water, repeating the performance until you have used up the ants. This last you must, of course, do before the ants begin on the honey, for they may prefer the honey to a sweetened sponge.

Honey, Kind to Produce.—Q. I have been running my apiary for chunk honey, but find that I can find a sale for quite a lot of extracted honey. I have a few nice, straight combs on medium brood-foundation, wired. What would be the storing capacity of one colony with 1-inch foundation-starters, one colony with full sheets of medium brood-foundation, as compared with a colony with full-drawn combs; that is, if a colony with full-drawn combs could fill 20 frames, about how much could the other respective colonies fill, everything else being equal? I expect to use full sheets of thin surplus for chunk honey, and full sheets of medium for extracting. I ask these questions simply to have some idea as to how much foundation of each kind to buy this season.

A. I don't know. If you want me to guess, I'm willing to do my best at guessing. I must premise that by saying that the answer depends somewhat upon the flow. If a short and very heavy flow is on, the fully drawn combs will have a much greater advantage than they will have in a light and long-continued flow. In the former case, while the colony with full combs stores 20 pounds, the colony with one-inch foundation-starters will store from 10 to 15 pounds, and the colony with full sheets of thin surplus from 12 to 17. With medium brood-foundation it ought to do just a little better than with thin surplus.

In the case of the long, slow flow, while built combs give 20,

the starters should give 15 to 18, and the full sheets of foundation
16 to 19. I can, however, imagine an extreme case with an im-
mensely heavy flow lasting only a day or two, in which 20 pounds
would be stored in built combs and not a drop in the others. On
the other hand, I can imagine a very long flow with a very little
more gathered daily than the bees need for their own use, and
very nearly as much stored with starters as with full combs. But
remember that all this is only guessing, and my guesser may not
work in perfect order. I think the editor-in-chief knows more
about it than I do, and I'd be glad to have his guess, even if it
makes mine look like the guess of a beginner.

(My guess would be a greater difference when built combs
yielded 20 pounds, say 10 to 15 pounds for starters, and 15 to 18
for sheets of foundation. I have seen sometimes what Dr. Miller
states, 15 to 20 pounds in built combs and not a drop in the others.—Editor.)

Honey, Purity of.—Q. .Some dealers tell me that I have been
feeding my bees sugar syrup. Others ask me if it is machine-
made. I would like to be able to prove that my nice, white comb
honey is pure honey, produced by the bees, but as I am not very
well posted on honey yet, I do not know just what to say. I have
heard it said that somebody, somewhere, offered $1,000 for a
pound of machine-made honey. Who was this man, and is the
offer still good, and has he got the $1,000 yet? The trouble is
that many persons believe that clean, white combs without stains
are machine-made; that pure amber honey is colored, and if it is
clear and white it must be nothing but sugar and water.

A. An argument that I think was first advanced by C. P.
Dadant ought to be enough to convince anyone with sufficient
reason that section honey is not machine made. Take any two
sections of honey and place them side by side. If machine made
they would be exactly alike; whereas there will be no difficulty in
pointing out differences that will knock out all idea that they are
made in the same mold, and establish clearly that each section is
an individual job, worked out by the bees. Pop-holes in one will
be clearly different from those in another, and variations of cells
will be evident. You may also show a section just as it is when
you give it to the bees, and that will be convincing to most men
that the bees do the rest.

The offer of $1,000 for a section of honey made without the aid
of bees was first made by the A. I. Root Company, and is still
good, with many thousands of dollars back of it. No one has yet

captured the reward. The same offer has also been made by the National Beekeepers' Association.

Honey, Ripe.—Q. When honey is sealed and capped over by the bees, is it ripe and ready to take off, if not, how is one to know?

A. As a rule, when honey is sealed it is ripe, and it isn't ripe until it is sealed. That's the rule, and if you follow it in taking off honey all the mistakes you make will never send you to the penitentiary. As with most rules, there are exceptions. The bees may seal up honey before it is ripe, and they may leave it unsealed after it is ripe. You can tell by seeing whether the honey is thick or thin. If it's thick, call it ripe. But the exceptions are so few that in actual practice I never paid any attention to them, merely counting all honey ready to take off if sealed.

Honey, Soil Affecting Yield.—Q. Why is it that some plants produce honey in some places and don't in others? Cotton, for instance, yields heavily in both north and south Georgia, but does not yield honey, or the bees do not get it, just a little north of the center of the state, among the red hills.

A. I don't know; only I know it is so. The soil or the elevation may have something to do with it.

Honey, Sour.—Q. (a) Will honey extracted from comb freshly built and not capped over sour if placed in a can? If so, how would you prevent this?

(b) Will comb and extracted honey put in regular honey buckets sour if kept any length of time?

A. (a) Maybe, and maybe not. Sometimes honey is sealed before it is ripened, but generally not. The remedy is to wait until the honey is sealed before extracting. Even if it never soured, it will be money in your pocket in the long run if you never put anything on the market but the very best ripened article.

(b) Either kind may be kept for years without souring if well ripened by the bees and then kept in a dry place where it will not attract moisture. Keep it in a place where salt will keep dry. If salt gets moist in a certain place so will honey, unless it be extracted honey tightly sealed.

Q. What makes honey sour in the hive when the flow is at its best and no honeydew? This season I ran my bees for comb honey; in some of the hives honey soured before it was capped.

A. I don't know. I know it sometimes occurs, and I suppose it is something in the character of the honey itself.

Q. Is it safe and right to feed bees honey that has soured?

A. Yes, if fed in the spring at a time when bees are flying daily, and at a time when there is no danger of its going into surplus. They will use it for breeding.

Honey, Thinning.—Q. How do you thin down honey to feed bees? If I mix it with water it sours.

A. Merely stir the honey and water together—all the better if the water is hot. It will not sour if fed within a day or so, and it ought not to be allowed to stand longer.

Honey, Unripe.—Q. What is "green" or "unripe" honey? Is not honey good to eat as soon as it is capped over?

A. Green or unripe honey is that which has not been in the hive long enough to become sufficiently evaporated. Generally it is sufficiently ripened when sealed, but there are exceptions. Fortunately, the exceptions are rare.

Honey-Board.—Q. How is a honey-board made? Is it just a board with a bee-escape in the underside? Would it pay to use these for extracting honey; put them on in the afternoon before you expect to extract and then just take off the supers and frames together, and wheel them into the house?

A. Honey-boards were in use long before bee-escapes were heard of. A honey-board was one placed over the top-bars, with a beespace between, there being in the board holes or slots, over which were placed surplus-boxes. Latterly a board with an escape in it is sometimes called a honey-board, but it is better to simply call it a bee-escape. Opinions are divided as to using bee-escapes in the way you mention, some highly approving them and others not believing them worth while. I suspect that bee-escapes work better for some than for others, either because of the difference in bees or for some other reason.

Honey Comb (See Combs.)

Honeydew.—Q. How can one tell when there is honeydew?

A. I don't believe I can tell you in words how you can decide as to honeydew. I couldn't tell in words just how an orange tastes. The dark color of honeydew and peculiar taste help to decide, but I can't tell what that taste is. If you know the bees are working on some tree where there are no flowers, you may be suspicious.

Q. Does honeydew ever appear before July 1st?

A. Yes, it does, sometimes.

Q. Does honeydew come any time of the year? My bees

seemed to be storing something in the warm days of February, before there was any blossoms of any kind. (Georgia.)

A. Honeydew may come almost any time plants are growing; but I supect your bees are working on something else than honeydew in February.

Q. Will bees work on honeydew during a flow from clover or basswood?

A. Not to any great extent. They prefer the better article of food.

Q. The grading rules of Colorado class honey contaminated by honeydew as not permitted in shipping grades. How is honeydew detected in the comb?

A. I'm not sure that Colorado officials have any particular rule as to how it is to be detected, but a good guess can be made by both looks and smell while in the comb, and if necessary it can be sampled by taste. It generally has a cloudy, dark look that honey does not have, and its smell is peculiar. Even if a certain sample of honey could not be positively identified as honeydew, if it were so much like it as to make it difficult to decide, I suppose it would be ruled out.

Q. My bees are working on honeydew, the trees just glistening with it; the leaves look as if they were varnished, and in the morning when the dew is on the bees work "to beat the band." I have several hundred pounds of it in the supers. It is bad-looking stuff and not fit to eat or sell. What can I do with it? Will it do to feed bees?

A. It will do to feed to the bees in the spring or any time when they will use it for brood-rearing; but don't give it to them for winter stores. Such honey may be sold for baking or mechanical purposes or it may be made into vinegar. It is also used by manufacturers of chewing tobacco.

Q. (a) Why is it that in honeydew seasons some colonies gather more honeydew than others? Such has been my experience.

(b) Do certain races gather less honeydew than others? I have been told so.

A. (a) Possibly there is a difference in colonies as to their preference for different sources. One year I had one or more colonies that gathered honey of light color while the rest gathered buckwheat. It might be that they strongly preferred the lighter honey, or it might be that they just happened on the lighter honey in some particular place.

(b) It is possible.

Q. Give your opinion as to bees wintering in cellar on honey-dew. My bees used up a large portion of this honeydew which they gathered earlier in the season in summer breeding, but, while they have a good quantity of honey to winter on, much of it seems to be dark and of a strong, almost sourish taste.

A. A small amount of honeydew in winter stores seems to do little or no harm, but in large quantities it is likely to do much harm. Of course, there is a difference in honeydew in taste, and there may be kinds not so bad for wintering as others, but it is not safe to count on that. Although the honey crop was a failure, it is just possible that you had a pretty fair fall flow, and that as the brood-nest became less the vacant cells were filled with honey of good quality for wintering. If you had extracted in September, or even early in October, and fed sugar syrup, it might have been safer. You might, however, lay a cake of candy, say an inch thick, over top-bars, if you think you dare not risk what is in the combs, as the bees would be likely to use the candy first.

Honey Locust.—Q. How does the honey locust compare with the linden in yielding nectar?

A. Not nearly so good. But it comes earlier, when it may help greatly in brood-rearing. The black locust is better.

Honey-Plants.—Q. I would be pleased to know if there are works on honey-plants. I have a couple of acres to devote to artificial pasture just for the honey, if it is probable that success might come of it in any way.

A. There is probably no work published that treats particularly of honey-plants, although the text-books on bee-culture give some information regarding them. It is not likely that you will find any plant that will yield sufficient honey to make it profitable for you to occupy land with it unless it yields a profit in some other way. Sweet clover will probably come as near it as anything you can find. If stock in your locality has learned to eat sweet clover, either green or dry, it will pay to occupy good land with it.

Horsemint.—Q. I have some horsemint plants on my place. Will they yield honey, and if so will this honey hurt the grade of my other honey?

A. Horsemint is not usually abundant enough in the northern sections of the United States to produce surplus quantities sufficient to make the taste noticeable. However, in Texas, horse-mint honey is a regular crop. I should say that if you had horse-mint in sufficient quantities it would probably give a mint flavor to your other honey. It yields about the same time as white clover.

Houses, Getting Bees From_ (See Buildings.)

Hybrids.—Q. Is the hybrid of yellow color?

A. When the word "hybrid" is used concerning bees, it generally means a cross between blacks and Italians, and such hybrids may have one, two, or even three yellow bands similar to the yellow bands of Italians, but if only part of the workers have the three bands, then the colony is considered hybrid. I suppose the word "hybrid" might also apply to a cross between Italians and Carniolans, or between any two different varieties.

Q. Are hybrids as good for honey-gathering as full-bloods?

A. Very often they are.

Q. Are hybrid bees as good as pure stock?

A. Sometimes they are better, and sometimes not so good. But even if better, they are more likely to run out than pure stock.

Hydrometer.—Q. On page 5 of the American Bee Journal for 1908, in an article on "Testing Honey as to Ripeness," it is said, "it would be a good thing" to "get a hydrometer." What is a hydrometer, and, especially, how is it used? Of course, I understand a "hydrometer" must be an instrument to measure moisture. Still, I repeat the question, what is it and how is it used?

A. A hydrometer is an instrument for determining the density of liquids, consisting of a weighted glass bulb with a long stem on which there is a graduated scale. It is put in the liquid, where it stands upright, the denser the liquid the higher it stands, the figures on the scale thus showing the density. If an up-to-date dairyman is near you, he may have a hydrometer, which he calls a lactometer.

Ice.—Q. If small particles of ice form on the hive-entrance of a colony that is wintering on the summer stand, is it an indication that it is not in a good condition for winter?

A. No, there is nothing alarming about it any more than there is in seeing a man's breath form in icicles on his beard when he is out in freezing weather.

Inbreeding.—Q. What do you call inbreeding? Give a practical illustration.

A. "Inbreed," says the dictionary, means "to breed or to follow a course of breeding, from nearly related animals, as those of the same parentage or pedigree; breed in-and-in." It would be inbreeding to have a young queen meet a drone from the same hive, or even with the relation less close.

Q. Does the honeybee degenerate through inbreeding? If so, what is the result?

A. Indiscriminate inbreeding among bees, as with all other animals, is likely to result in deterioration, the bad qualities becoming intensified. With intelligent control the result may be the other way.

Income From Bees (See Living From Bees.)

Increase.—Q. I am 21 years old, and I own four colonies of bees. I am as interested in the bee-business as I think any person can be. I have read all the bee literature I could for three years. At present I am taking four bee-papers. Would you advise me to buy more bees, or to wait until those I have increase?

A. That depends. If you want to increase to a considerably larger number and have an opportunity of buying a few colonies at a bargain, as sometimes happens at an auction, or when one wants to get rid of his bees, it will be well for you to buy. But if you can't buy for less than $5.00 a colony, then it will be more profitable for you to run your bees for increase than for honey. Only don't make the mistake of having a number of weak colonies on hand in the fall. It would, no doubt, be an easy thing to increase those four colonies to twenty or more by fall, and then lose most of them in the winter because too weak; but in the long run you will get on faster to move a little more slowly and surely. Of course, something depends upon the season. In a very poor season it may not be safe to increase at all, unless you do a good deal of feeding. But if you reach next fall with ten or twelve strong colonies, another good season ought to bring you up to forty or so.

Q. Can you make a 20 per cent increase by going through the apiary and making a colony at different times without hurting the honey-flow?

A. I think it might be done without diminishing the crop, at least in some cases. Just enough strength taken from each colony to prevent swarming might increase rather than diminish the total harvest.

Q. Can I take a colony and make four or five out of it and put a new queen in each? If so, how?

A. You may do it in a good season. One way is to wait until the colony is strong, then take a little more than half the brood and bees and put in a new hive on a new stand, giving a new queen and leaving the old queen on the old stand. When each of these becomes strong, divide again the same way.

Q. As I am only 25 miles from you, please recommend the best method to increase and still get a crop of honey, for our locality. I have your book "Fifty Years Among the Bees."

A. There are so many different circumstances and conditions that it is not easy to say what system is best. What is best one time may not be best another. In the book you mention the matter of increase is discussed as fully, at least, as in any book I know of. After a careful study of what you find there, you will be able to decide for yourself better than I could decide for you. If, however, I were obliged to confine myself to any one plan, with the idea of interfering little with the honey crop, I think it would be the nucleus plan. With that you can make much or little increase, and you need not draw from one colony enough to hinder it from doing fair work in supers. But if by "still get a crop of honey" you mean to get as much as if you got no increase, I don't believe you can make it in your location. That only happens where there is an important fall flow.

Q. Is the Swarthmore method, i. e., shaking the bees on full sheets of foundation and then giving them a laying queen, better than the Alexander method of increase, as in "A, B, C of Bee Culture?"

A. Likely the Alexander plan may be better for you, as it allows little or no chance for brood to be chilled. But if you expect to double your crop of honey, as Mr. Alexander says you may, by dividing, you are likely to be seriously disappointed unless you have a heavy late flow, as Mr. Alexander had from the buckwheat.

Increase, Alexander Plan.—Q. What is the best way to double any number of colonies?

A. Something depends upon circumstances what is the best way. If you have had very little experience it may be best for you to depend upon natural swarming, but allowing no afterswarms. When a colony swarms, set the swarm on the old stand and set the old hive close beside the new one. Then a week later move the old one to a new place ten feet or more distant. That will prevent afterswarms, and the swarms will give you surplus if there is any surplus.

If you prefer not to have natural swarming the Alexander plan of increase may suit you. A little before it is time for bees to swarm in your neighborhood, lift out of the hive all but one frame and put them in an empty hive-body. Leave the queen with the one frame of brood, and destroy any queen-cells that may be on that frame. Fill out both hives with frames filled with foundation, or with starters or drawn comb. Put a queen-excluder over the hive containing the queen and one brood-frame, and set the other hive on top of this. Five days later look for queen-cells in the upper stories. If you find queen-cells in an upper story, let it stand another five days, and then set it on its new stand, giving it a queen-cell from one of the others. It will hurry up matters if you can give a laying queen to each.

Q. When taking the top story off, how many bees should go with it?

A. I think Mr. Alexander took all that were in it.

Q. Do you consider the above method better than allowing natural swarming with clipped queen, or dividing by forming nuclei?

A. No, not for me, and probably not for one in a thousand in the North.

Q. I have three colonies and should like to increase and also try Caucasians. Could I take one or two frames from each colony, unite them and then introduce a Caucasian queen? Will it prevent the mother colonies from swarming? Can you suggest a better plan if mine isn't practicable?

A. Yes, your plan is feasible. But taking away only one or two frames of brood from each colony is not likely to prevent swarming, although it will delay, and in a few cases prevent it. To fulfill your desire you will do well to follow what is called the Alexander plan, varied a little. Wait until the time comes when there is danger of swarming. Then put all brood but one frame in a second story, leaving in the lower story the one brood and queen, filling out with drawn combs or frames filled with foundation, and pay no attention to where the bees are. Put a frame of comb or foundation in the second story to fill out the vacancy. Have a queen-excluder between the first and second story. A week or ten days later lift off the second story and set it on a new stand, destroying all queen-cells, if there are any. Twenty-four hours later give to this new colony a laying queen, a virgin, or a queen-cell.

Q. I have two strong colonies of bees; in each hive the brood-chamber is a double 10-frame brood-chamber, making 20 frames to each.

Now, I wish to know what is the best way to make "increase"

of my bees. I would like to avoid the troubles of the usual swarming, and yet increase my stock.

A. Here is one good way: Operate a little before the usual time of swarming in your neighborhood; or, if you wish to take a little more pains, operate after queen-cells are started, but before they are sealed, for with the first sealing a swarm is likely to issue. Set one of the stories on a new stand, putting in it all the frames of brood with adhering bees and leave the rest of the combs and bees with the old queen on the old stand. The hive on the old stand ought to give a good surplus in a good year. There is, however, some danger of a swarm as soon as the first young queen emerges. You can prevent this by destroying all queen-cells but one. Or, you may prevent it by dividing the brood into two parts, providing you want the increase.

It may be still better first to put all the brood in the upper story, with an excluder between the two stories, and the queen in the lower story. Then, a week later, move the upper story to a new stand. In this case there ought to be no danger of swarming.

Q. Referring to the Alexander method of preparing colonies for division, by placing the older brood above an excluder until sealed, and the queen with open brood below upon the bottom; I would like to divide as early as there is sufficient brood, and, for that reason would like to know if the process might not successfully be reversed, the queen being placed above the excluder, and the brood for sealing below, and thus avoid desertion of the queen by reason of unexpected cold, which has been reported by one observer.

A. Yes, the queen may be put above the excluder, leaving the brood below, although you will probably not like it quite so well, for the natural thing is for bees to work downward, with the brood.

Increase, Artificial, by Division.—Q. I have 28 strong colonies and want to increase to 50 if possible this season, and would like to do it artificially, as I think it will save a lot of time. This is my second season with bees. How shall I proceed? Would it do to divide the bees just before they are ready to swarm, and is it best to put frames of foundation in the old colony where I take out the frames of brood?

A. Yes, one of the simplest ways is to divide each colony into two parts before the bees swarm. Leave the old queen on the old stand and put more than half of the brood with adhering bees on a new stand and they will rear a queen. Fill vacancies with frames having full sheets of foundation.

But that's far from the best way. Just what the best way is, depends upon circumstances, and it would take much room to go fully into the whole subject of artificial increase. Study up general principles in your text-book, and you will be better prepared to judge what is best for you.

Q. Which is the better, natural swarming, or dividing?

A. Whether the swarm made by dividing is as good as a natural swarm depends upon how the natural swarm is made. It may be made weaker than a natural swarm, and it may be made stronger. There are, however, advantages in dividing such that experienced beekeepers generally prefer it to natural swarming.

Q. How late would it be safe to divide, and also to buy a queen? Our seasons are long here. The 10th of October is a very early frost. Cotton blooms till frost. (Oklahoma.)

A. I don't know just how late it might be safe to divide. If the flow continues until October 10, and a laying queen is furnished, you might risk a division as late as September 1 to 10, providing the colony be strong, with plenty of brood.

Q. How many days after the swarm issues should I divide?

A. About seven or eight days.

Q. When is the best time to make new swarms? I worked with a beeman one summer before I bought my bees. He made his new swarms when he was extracting. But I think it disturbs the bees so much when they are working hard, and it looks to me like they will not store as much honey if torn apart at this time.

A. There is no fixed rule about it. One would think it best to follow nature and make increase at the time bees swarm naturally. But nearly everyone agrees nowadays that natural swarming is decidedly detrimental to the honey crop. In my locality it seems much better to have no increase until at or near the close of the harvest. In some localities, where there is a heavy late flow, it may be better to divide early in the season.

Q. When is the proper time to start new colonies of bees in this climate—40 miles south of St. Louis? What is the best method for a beginner to take in doing so?

A. The very best time is when the bees are inclined to swarm naturally. Bees begin to swarm when honey begins to yield well, and more or less colonies may swarm so long as honey yields, although most colonies do their swarming during the early part of the honey-flow. You may even make increase successfully in the month of September if you make the new colony strong enough.

The earlier you start a colony the less need of its being strong, as it has a longer time to build up before winter.

It is not easy to say what may be the best way for you. What may be best for one may not be the best for another. Perhaps the easiest way is to take half the combs, bees and all, out of a hive and put into another hive, filling out each hive with combs or frames filled with foundation, setting the hives side by side, as nearly as possible on the old location, trusting to the queenless part to rear its own queen.

A better way is to look four days later and see which hive contains eggs, and give a laying queen to the other part. That, of course, involves buying and introducing a queen.

If you want the bees to rear their own queen, here is a better plan: Find the queen and put her with two frames and all adhering bees into another hive on a new stand. A week later a number of queen-cells will be in the now queenless colony, when you let the hives exchange places, and the bees will do the rest. If you want to have more than one new colony, you can divide the queenless part, putting the larger half on an entirely new stand.

Increase, Artificial, Nucleus Plan.—Q. Give me a good way for artificial increase.

A. Take from the strong colony two frames of brood with adhering bees and queen, put on a new stand and imprison them for three days. A week after the queen is taken away, let the two hives swap places. That will double the number. If you want to make more out of that one colony, you can divide the old colony into two or more nuclei at the time of swapping places, being sure that each has a good queen-cell located centrally where the bees will keep it warm, and then if necessary you can strengthen these nuclei after the queens get to laying by giving them brood from other colonies.

Q. I have a colony of Italian bees from which I intend to make increase. If I make nuclei from it, will it be safe to give frames of brood with adhering bees from other colonies? Or is there danger that the bees will kill the queen or destroy the queen-cells? If this is not safe, how fast can frames of brood without bees be given? I understand if too much brood is given at once some will starve.

A. It requires judgment in giving frames of brood with adhering bees, as it depends upon the strength of the nucleus how much can be given at a time. You evidently have in mind two dangers. One is that the strange bees introduced will kill the queen (there is not very much danger that they will harm the queen-cells); and the other is that the brood will be chilled or starved. Generally more danger of chilling than starving.

Unless a nucleus has bees enough to cover three frames, it is better not to introduce a frame of brood with adhering bees, lest the queen be endangered. With regard to brood, there is little danger of harm being done if bees enough go with it to cover an additional frame.

In any case, the more mature the brood the better, and if the brood is all sealed you may give a frame without any adhering bees, and it will be safe in a nucleus of two or three frames, even if there appear to be only enough bees present to cover well the two or three frames already present. One reason for this is in the fact that it does not require so much heat for sealed as for unsealed brood. As soon as most of the young bees have emerged from the frames given, it can be exchanged for another, and this will generally allow you to add a frame each week.

A nice way to do to have frames of brood ready to give to nuclei is to put an excluder over a strong colony with an empty hive-body over it, and put into this frames of brood from other colonies; then, a week or ten days later, there being no young brood present, the frames will be fine for nuclei, whether you take with them the adhering bees or not.

Q. Being anxious to increase as fast as possible, I would like to have your opinion about it. I read one article by W. Z. Hutchinson, saying that he made his increase by taking two or three frames of brood from strong colonies and giving them a laying queen; but not being able to buy my queens, would it do to take a queen from one colony and let the bees rear a new queen? Please give me some of your best plans.

A. It is not easy to say what may be the best plan for you. What is best for one is not always best for another. But taking the plan you mention, you can do very well with a little variation. Decide which colony you think has the queen of best blood, and see that it is strong, if necessary giving it frames of hatching-brood from other colonies to strengthen it. You may even fill two stories with brood. Call this hive A. When the time comes for bees to begin making preparations for swarming, take the queen with 2 or 3 frames of brood and adhering bees, and put them in hive B, on a stand a rod or more distant. About eight or ten days after taking the queen away from A—don't delay longer than ten days—take out one of the frames with the queen from B, put it in an empty hive, C, and fill out C with empty combs or frames filled with foundation-starters. Take hive A from its stand and set hive C in its place. You will now make as many nuclei as you can from the brood and bees in A, taking two frames of brood and bees for each nucleus, putting each on a new stand. It may happen that without any intention on your part there will be one or more good queen-cells on one of the combs in each nucleus. It may be, however, that most of the queen-cells are on one or two combs, and you must cut out at least one good cell for each nucleus. You can fasten it on the comb by pushing over it a hive-staple. See that it is centrally located where the bees will be sure to keep it warm. A cell must also be given to B, and it will be well that this be given in a cage so that the bees cannot get at it for a day or two, lest they destroy it before they discover their queenlessness. The bees of the nuclei being queenless, will remain pretty well where put, but you might fasten them in for a day or two.

Increase, Natural.—Q. I have 25 colonies of bees and want to increase to 50 next year and secure as much surplus honey as possible. How would you do this? We have plenty of white clover that begins blooming May 1, and blooms two months.

A. There is, perhaps, no better way than to let each colony swarm once, moving the parent colony to a new location and hiving the swarm on the old stand. That will give a strong force to the swarm, which will do the principal storing, although the mother colony may store some if there is a late flow.

Increase With Queen-Cells.—Q. Is it safe to form a colony by taking frames of sealed brood and queen-cells instead of queen?

A. Very unsafe if no precaution whatever is used. For when you look a day or two later you are likely to find the bees mostly gone and the brood chilled. After putting the two frames with adhering bees in your nucleus hive, shake into it the bees from one or two more frames. Then see that your hive is closed beetight, so that not a bee can get out, for two or three days. It's not a bad plan to stuff grass or green leaves into the entrance, plugging it tight and hard. The green stuff will dry and shrink, and in two or three days the bees will dig their way out.

(If the weather is very hot, better keep this hive in the cellar during that time, as they might smother in the hive under the sun.—C. P. D.)

Increase, Prevention of, (See Swarm Prevention.)

Introducing (See Queen Introduction.)

Italian Bees.—Q. Has Italy two kinds of Italian bees, the leather-colored and the golden; or are the goldens bred in this country by select leather-colored stock?

A. In Italy there are the leather-colored and also a lighter kind, but I think no 5-banded or golden, which is an American affair, not at all always from the leather-colored kind.

Q. How can I tell a pure-bred Italian queen? I notice all the queens I buy, and also the drones, vary in markings.

A. The workers should not show less than three yellow bands. But you may find in a colony of pure Italians, black workers that have come from other hives. Look for the downy little chaps that are quite young; amongst them there should be none without three bands.

Q. Are Italians the only bees having three yellow rings on the abdomen? Should the rings be wide or narrow?

A. There are others having three such rings, as the Cyprians. It doesn't matter about width of rings, the distance of one ring from another being the same in all cases.

Q. Please distinguish between leather-colored Italians and other Italians.

A. Leather-colored Italians are, as the name indicates, rather dark in color, the colored part being the color of sole-leather, as compared with other Italians of lighter color.

Q. I would be glad to know what the difference is between the 3-banded and the golden Italian bees, and how are they obtained?

Also, are the 3-banded bees longer-tongued than the golden? I had goldens that worked on red clover, but I see they are always classed differently.

A. The workers of bees imported from Italy have three yellow bands. Those that are called golden are obtained by breeding continuously from the yellow races, constantly selecting those showing most color. They are an American product.

There is probably no difference as to length of tongue between the two classes. When bees work on red clover it may be because of longer tongues, and it may be because of shorter corollas in the blossoms. I have seen black bees working on red clover.

Q. Which stock is best to order, the 3-band or 5-band?

A. Some prefer those with more than three bands, but most

prefer those which have three bands, like the pure stock that comes from Italy.

Q. Are the 5-band, or golden, any better than, or as good as, the 3-band leather-colored Italian queens?

A. There are goldens and goldens. Some are good, and some are poor, according to all accounts; while the 3-banders, as imported from Italy, are more uniform and of a more fixed type.

The beginner is generally puzzled to know whether to choose goldens, bright 3-banded, or leather-colored. Let it be distinctly understood that all goldens are not exactly alike, neither are all leather-colored. The three kinds mentioned are all Italians, and they all vary. So a man may have a colony of goldens and a colony of leather-colored, and the goldens are the better of the two, while another may find that his own goldens are not so good as his leather-colored. The matter of looks has no small bearing, and breeders find that, other things being equal, the brighter the color, the better the customers will be pleased. Yet a large proportion of experienced producers of honey seem to prefer the leather-colored, with the belief that in general these rank as the better honey-gatherers.

Q.. Are the goldens generally recognized as the worst robbers of all bee-kind? The ones I have certainly must be; however, with the miserable slow flow we are having, they are certainly getting much more honey than are my blacks.

A. I don't remember to have heard that charge against the goldens. I am afraid it's true very often that the best gatherers are inclined to be bad as robbers. Bees have no moral sense, and don't make any distinction between getting stores from the field or from another hive; so why shouldn't the best gatherers be the best—or the worst—robbers?

Q. Do you find the yellow Italians more vicious than those of a darker color?

A. They vary; some are vicious, and some gentle.

Italianizing.—Q. What is the best time of the year to Italianize?

A. Other things being equal, there is no better time than toward the close of harvest, but if I had poor bees I wouldn't want to wait till then to get better stock.

Q. I have a few colonies of black bees that seem to be weak, though they are beginning to carry pollen. Would it not be better to wait until later in the season before I attempt to introduce Italian queens. Should I not catch the old queen and destroy her before I send for the new one? I have heard there is less danger of having the queen killed when she is introduced if the colony has been without a queen for some time. Should the queen be clipped before she is introduced?

A. You will probably do as well to wait until some time in June. Better not kill the old queen till the new one arrives. There may be considerable delay, and it is not well for the colony to be too long queenless. You can have the same, or greater, advantage by keeping the new queen caged in the hive two or three days before allowing the bees of the colony access to the candy to liberate her. Most beekeepers nowadays prefer to have queens clipped, and most of those who sell queens will clip them before sending, without extra charge, if you so request.

Jouncer.—Q. What is a jouncer?

A. A frame-work upon which a super rests, allowing the bees to be shaken out of the super by jouncing the ends of the jouncer up and down alternately. It has not proven a success with everyone.

Krainer Bees.—Q. Have the Krainer bees from Krain, Austria, ever been imported to this country? Are they more hardy than Italians?

A. You have probably heard and read quite a little about Carniolian bees. Well, Krain is merely the German word for Carniola. I'm not sure whether the claim for greater hardiness has been well established, but some think well of a cross with Italians.

Labels.—Q. How do you stick labels on tin cans? I don't seem able to make them stick.

A. The favorite way is to have the label pass clear around the can and overlap, in which case any common flour paste will answer. Flour paste with water sticks to tin. Wiping the can off with a dry cloth to remove the greasy substance left in tinning will help make labels stick.

Larvæ.—Q. Is there any way of determining accurately the age of the larva, or, in other words, how long it has been in process of development, without waiting for it to be sealed over?

A. Nothing very definite. In general terms it may be said that the larva makes most of its growth in the last two days of its five days of larval existence; and I think it doesn't cover the bottom of the cell till after it is three days old.

Q. At what time and in what way are the young bees fed?

A. They are fed by the nurse-bees for five days or more from the time the larva hatches out of the egg until it is sealed over.

Laws on Beekeeping.—Q. What are the laws governing the keeping of bees, disease control, etc.?

A. Laws concerning bees and diseases are made by each state individually. Write to your state bee inspector or state entomologist for information. If you do not know who he is, the publishers of the bee journals should be able to inform you.

Laying Workers.—Q. In overhauling a friend's bees today, I had a new experience. I found a colony that was queenless (at least no queen was noticed), and no brood was found in any of the combs, excepting a small quantity of drone-brood sealed up. Ninety-six drone-cells, actual count, and one sealed queen-cell; no worker-brood at all. The queen-cell was right among the sealed drone-cells.

A. It is not only possible, but probable, that not only a laying worker, but a number of them were present, and that the bees attempted to rear a queen just as you have suggested.

Q. In appearance are laying workers different from workers?

A. Not a bit. I suppose I have seen hundreds of them—for in every colony with laying workers it isn't a single worker, but a whole nest of them at the miserable business—but I never could tell which the laying workers were, except one single laying worker that I caught in the act of laying.

Q. Do old bees become drone-layers, or do only the younger ones "go astray?"

A. I have a strong impression it's only the younger ones. Some have advanced the theory that laying workers, in their larval existence, have been located near queen-cells, and so have been fed some of the royal jelly as a sort of overflow. If that were the true theory, of course there would be no drone-layers except those which started in at the business early in life. But I wouldn't take much stock in that theory. Nurse-bees are not so careless as to slop around the soup dishes in that sort of style. Besides, if that theory were correct, laying workers would be just as likely to appear at all times after young queens are reared, whereas we know that with most races of bees no laying workers are seen unless a colony has been hopelessly queenless for some time. I don't remember that I ever saw any other explanation given, but if you can't find anything better I'll offer one of my own for what it is worth. It is that when a lot of nurses are loaded up with pap, and only a few larvæ are left unsealed, those few are fed so heavily that they are developed sufficiently to do something in the egg-laying line. If any reliance can be put upon this theory it is still true that no bee could start in as a laying worker after it becomes old.

Q. Different writers claim that drone-laying workers are the only ones guilty of laying eggs on the sides of a cell. Last fall I found a colony with a drone-laying queen of previous year's rearing, and I found lots of worker-cells with two or three eggs in a cell, some at the bottom and others stuck to the sides half way down. In such a case, is the colony liable to have laying workers acting in conjunction with the drone-laying queen?

A. I think I never heard of laying workers being present with a laying queen, at least for any considerable time. Queens sometimes lay eggs on the sides of cells.

Q. How am I to get rid of a laying worker?

A. Generally the best thing to do with a colony that has laying workers is to break it up, giving the bees to other colonies. It is difficult to get the bees to accept a queen. But if the colony is strong enough, and you are anxious to have it continue, you can give it a virgin just hatched, and this will pretty surely be accepted. Or, you may exchange some or all of its combs with adhering bees for frames of brood and bees from another colony or colonies, and the younger bees thus introduced will accept a laying queen.

Q. Did you ever have any experience with laying workers in a hive where a young queen has hatched? This is my experience: On May 9 I transferred a swarm of bees from a hive which I expected to discard on account of its odd size. On May 30 all the brood was hatched, and on examination I found a queen-cell already hatched, and by searching I found the young queen. Today I went through the hive to see if the queen was laying, and all of the eggs and larvæ which I found were in drone-cells and the eggs scattered about in worker-cells. I examined closely the comb on which I found the most drone-cells, and then and there I saw a worker doing her work. What do you think of that, with a young queen in the hive, and she a beauty? I closed the hive, thinking things might right themselves if left alone, but in the afternoon I found the queen on the alighting-board dead, with a ball of bees about her. I broke up the colony at once. Would you kindly tell me what you think of this case? When I say they had a laying worker, I mean to say that I saw her lay one of her eggs in a drone-cell.

A. Your experience is quite exceptional. It is not often that a laying worker is caught in the act. In all my experience I never saw it, I think, more than once. If your bees are Italians, it is remarkable that laying workers should appear when they did, although with some of the other races laying workers are inclined to put in an appearance whenever laying is not normal. You speak a little as if there were only one laying worker present. The probability is that there was quite a large number.

Q. You say a colony of laying workers should be broken up and the combs distributed among other colonies, and that the bees are old and of little value. In what way would it be doing any good to give a good colony one of those combs of worthless bees and drone-brood?

A. While these old bees are of little value, they still have some value, and that value may as well be utilized. We are told that a worker in the busy season lives to be about six weeks old. Now suppose we have some bees that are four or five weeks old. They have yet a week or two to live, and they are good as field bees for that length of time; so if given to other colonies they will finish up their lives in a useful way, doing more good than if you try to tinker up the colony with a young queen. To be sure, you might give a queen, together with brood, and enough young bees to make a fair colony, but these old bees are exceedingly loth to accept a queen, and you'll be likely to lose her. Better break up the colony, and then start a new one elsewhere.

Lazy Colonies.—Q. Last season I had lazy colonies that did not do anything but rear bees. They were running over with bees but did not swarm nor store any surplus honey. Would it be best to give them another queen?

A. It is possible that there may have been some excuse for the bees doing nothing, but if other colonies were doing well at the same time the likelihood is that the bees were at fault, in which case it would be well to give a queen of better stock. They may have been late in breeding, owing to spring weakness, and in such case the hatched bees, coming too late for the crop, would only help consume the honey harvested.

Q. Do bees "lay off" for a week before swarming and do nothing but eat honey?

A. They take no such vacation as you suggest. There may be a let-up for some hours, and you may see bees laden with pollen, among the swarming bees.

Q. Is there any way to get backward bees to work in the super besides baiting them, and then maybe wait until they are forced to? That was always my luck.

A. Yes, get them so strong early that they'll be glad to rush into the sections without any bait; only they will enter the supers sooner with baits. If you mean a way to make a weak colony start work in a section that will not begin on a bait, there is no such way. You may force them to go into the super by putting some brood in it, but that will not force them to store there if there is plenty of room to store in the brood-chamber.

Legs of Bees.—Q. I think I have something new this season. My bees have great long thongs dangling about their feet, and when they alight these thongs lie on the alighting-board to one side of the bees' feet. They are about one-eighth of an inch long, and just as red as can be. What are those false thongs on my bees' feet? Are they natural or not?

A. I think what you call "false thongs" must be the pollen masses from milkweed. In some cases it gets so bad that the bees can hardly climb upon the combs, and I have seen the other bees drive them out of the hive. Sometimes the bees are fastened upon the blossoms, not being able to tear away, and if you examine the milkweed blossoms in your vicinity you may find some dead bees upon them. But these plants are good honey-plants, and perhaps in this way pay for the injury done to the bees.

Leaves for Cushions.—Q. Are dry forest leaves as good as anything for the absorbent cushion?

A. Nothing could be much better, unless it be cork chips.

Lemon Juice.—Q. One man says to put lemon juice in sugar syrup and it is as good as honey for the bees. What do you think of it?

A. It would be better without the lemon juice, unless it be just enough to keep the syrup from granulating.

Lettering in Comb Honey.—Q. How is the lettering or spelling in raised letters on comb honey done? Please explain.

A. I know little about it practically. Usually this is done by inserting in the section or comb the letters in wood, then place these over a very strong colony to draw out and seal.

Lice.—Q. There is a spider-like insect that clings to the backs of my bees, in some cases covering the fore part of queens all over; of a reddish color, about the size of a pinhead. Is this the Italian bee-louse? It makes no difference whether the colony is strong or weak. Is there any way of getting rid of these? The temperature gets very high here in summer. Would the heat breed these things? (British Columbia.)

A. Yes, it may be braula coeca, or the blind louse, although I don't remember to have heard of such a case this side of the ocean. It is said smoking with tobacco will make the louse loose its hold. Fortunately, it is not so very dangerous. Please send a sample to Dr. E. F. Phillips, Washington, D. C.

Q. Will hen lice bother bees if one has them in a house apiary in the second story of a chicken establishment?

A I have never heard of bees being bothered by such lice.

Location.—Q. I would like your advice on choosing a location, as I am just beginning with bees and want to get the right location so that I may make the most honey. Some beekeepers report large crops. Should I locate in their section?

A. A young man just making a start, and intending to make beekeeping the chief business of his life, will do well to look about and choose a location specially suited to that business. For that matter, so might one already engaged in the business. But he would be unwise, especially if already well established, to go some distance to a new place, knowing nothing about it, but that some one had secured a very large yield of honey. Possibly that one year of great flow may be offset by two of failure. Possibly the distance or inconvenience of access to a market may be so great as to counterbalance the greatness of the yield.

Then there are considerations outside of beekeeping not lightly to be ignored. Climate, home and surrounding, are all of importance. Some of the northern beekeepers in attendance at the National meeting in Los Angeles some years ago, who had cast longing looks towards that golden land, went home entirely satisfied to remain where they were, after seeing some of the California apiaries. Of course, all locations in California are not the same, but some of them are dreary enough. To get the advantage of pasturage, an apiary is located in some canyon, away from the haunts of man, the nearest neighbor a mile or more away, outside of the sound of bell of church or school. With many it is a life of exile during the honey season, the rest of the year being passed elsewhere. But all would not like that sort of life.

On the whole, taking into account the ties of friendship, and old associations, as well as the trouble and expense of making a change, the probability is that not one beekeeper in ten will find himself better off anywhere in the world than right where he is now.

Q. If you were to start again from the beginning, intending to make beekeeping your life calling, and had no ties to bind you to any particular locality, where would you be likely to settle?

A. I would do a lot of investigating before settling. What might suit me might not suit you. I'm growing more and more to think that there's a good deal of equality in locations, advantages and disadvantages. I used to envy Californians. I'm not sure I would care to be there now, since knowing more fully about it. In general, the place where one happens to be is "not so worse" as it might be.

Q. How far from the public highway does the law require an apiary to be to insure one from all damages?

A. That depends altogether upon the local or state laws. Generally, I think, there is no law about it, but if you are wise you will not risk having your bees close enough to the highway to endanger anyone passing by. If your bees are very gentle it may be safe to have them close to the roadside. Some bees are not safe ten rods away.

Q. As I want to move my bees this spring, would two feet apart be too close for each hive?

A. That depends. If there are plenty of trees or other objects to help mark their locations it will be all right. If the ground is perfectly level, and nothing to help locate the hives, there will be mistakes in entering hives. If you want to save room, instead of putting them regularly two feet apart, put the first two close together, leave a space of three feet, then two more hives, and so on, putting the hives in pairs, with three feet between each two pairs. With that arrangement you'll get more bees on the same ground, and at the same time there will be less mixing.

Q. I have seven colonies of bees located in a valley, and a neighbor beekeeper has his bees on a high hill, perhaps 175 feet higher than mine. He says my bees have the advantage of his, as my bees go up hill empty and down hill loaded, while with his it is just the other way.

That is all right and true, as long as my bees go west or north, but when they go east they would have to go up first and then down on the other side, and I notice that they don't go very far that way. Now if I would place my bees on top of this hill I think they would go farther south and east than they do now. Most of the basswood is on the hillside, but the best clover is in the lowlands. Do you think it would pay to move my bees higher up for this reason?

A. It surely must be easier for a bee or anybody else to carry a load down hill than up. In actual practice I have some doubt whether the difference is enough so that a colony in the lower location would show a distinctly larger yield than one higher up. But the matter of distance may be a much more important factor. Within a distance of perhaps a mile and a half it is doubtful that distance counts for much. Beyond that—possibly I should say two miles—distance counts for a good deal, and if your source of supply is beyond that there will no doubt be a gain to move the bees accordingly.

Lumber for Hives.—Q. Is there any special lumber which should not be used in hives?

A. Basswood is bad, owing to its tendency to twist and warp. White pine is generally used, and also redwood, cypress and cedar.

Maples.—Q. What do you think of a locality from 1,000 to 2,000 feet above sea level, where there is a large quantity of maple sugar produced every year? Would you consider it a good locality for beekeeping?

A. The maple is a valuable honey-tree. It comes early, however, and the honey secured from it is mostly used in brood-rearing. The field-force is not yet strong enough to gather much more than will supply the daily needs of the colony. So while it is of value in securing a strong force of bees, the question whether the locality is a good one depends upon what comes later. If there are plenty of later sources the maple will be a great help; if nothing comes after, there is little prospect of surplus.

Q. Do bees get much pollen from the "sugar" or hard maple? What color is it?

A. They do in this locality It might be called light yellowish, with a tinge of green in it; possibly more green than yellow.

Q. How would it do to draw sap from maple trees in buckets and feed to the bees in early spring to start brood-rearing? Would this not be better than sugar syrup?

A. I don't know that maple-sugar syrup is any better than cane-sugar syrup. Likely, however, it is just as good. Care should be taken about feeding it when it is not warm enough for bees to fly.

Q. I have some maple sugar that has been damp so it is unfit for market. Would this make good food for bees?

A. It may be profitably fed next spring after bees are flying, to be used up in rearing brood; but don't give it to the bees for winter food.

Martins (See Bee Martins.)

May Disease.—Q. I have one colony of Italian bees in my yard that is dying from some cause. The adult bees are dying by the hundreds. They come dragging out of the hive, and sometimes crawl part way up the front; others fall off the runboard. They are trembling or jerking and moving their wings. Sometimes they just turn around in a very small circle, and sometimes they lie on the ground for two days, kicking or moving their legs until they finally die. Some have greatly enlarged abdomens, almost as large as a young queen, while others look shiny like they had just crawled out of grease or syrup. They have considerable honey and brood, but do not gain any.

(a) Do you think this is what they call May disease?
(b) What is the cause?
(c) Is it contagious?
(d) Do you think it would do any good to requeen?
(e) What can I do to get rid of this disease?

A. (a) Your excellent description marks the disease quite clearly as the disease called in Europe Mal de Mai, or Maikranheit (May disease), and generally called in this country bee paralysis.

(b) It is believed to be due to a microbe called bacillus gaytoni, and also bacillus depilis.

(c) It is not considered contagious; yet sometimes affects a whole apiary.

(d) I don't know. Some have claimed that as a cure; others say it does no good.

(e) I don't know. Many cures have been reported successful, only to fail when tried further, and as the disease has a way of disappearing of its own accord, the supposed cures may have no effect.

O. O. Poppleton sprinkles sulphur on the bees and combs. This destroys the sick bees, but it also destroys the unsealed larvæ, unless the brood be removed.

Mesquite.—Q. Where is the mesquite found, and what is its value?

A. The mesquite, a low, brushy tree, is found in Texas, New Mexico and Arizona. It yields honey best during very dry weather. In Texas they have a crop of mesquite honey nearly every year. The honey is amber in color.

Mice.—Q. I bought my partner's share of the bees, and on opening the hive I found a mouse-nest in it. I thought that very strange, having never heard of it before. Have you any mice in your beehives? The colony is a strong one, and I thought the bees would keep the mice out. It never destroyed any comb while in there.

A. Yes, indeed; I've had mice in hives, and they have not always been as considerate as yours, for they have sometimes gnawed the combs. You can keep them out by having the entrance closed with wire-cloth having three meshes to the inch. That will bar mice, but allow bees to pass.

Milkweed.—Q. (a) I enclose what we call milkweed. The bees work on it hard. A magnifying glass shows a sticky substance on it. Is it a good honey-plant?

(b) Why do bees stick to it?

A. (a) Yes, except for the trouble you mention in the next question.

(b) The pollen masses get fastened to their feet and stick so tight that the bees pull them from the plant and carry them away. (See Legs of Bees.)

Miller Cage.—Q. Would not a block 1¼ inches square with a ⅜-inch hole bored in the center answer as well as if I used two pieces ¼x⅛, a piece of tin and a piece of section each ¼-inch square? This is for a Miller cage.

A. It would answer just as well, except that it would take up too much room to be put between two combs.

Moisture in Hive.—Q. I find a dampness against the top of the hive, in winter. Should the bees have top ventilation, or will the dampness not do any harm?

A. If there is enough dampness so it will fall in drops on the bees it will do harm. Give upward ventilation by placing a cushion over the top to keep it warm. Then the dampness will not settle on them.

Q. I am wintering my bees out of doors in the following manner; take off cover, place a piece of burlap over the frames, place empty comb honey super on top of this, fill with chaff, put cover on top, raised one-sixteenth of an inch. I do this to prevent moisture from collecting inside. This does prevent it to a great extent, but even with this protection, when we had a spell of weather with the thermometer down to zero every morning for a week, some frost will collect on the walls and outside frames of the hives, and there will be some ice inside around the entrance; but the clusters are apparently dry and comfortable. Do you think this much moisture will keep the bees from wintering perfectly? I have tried packing outside on top of the cover. It didn't do much good.

A. When you go out in very cold weather on a long drive, you often find frost and ice collecting on the wrappings about the face. That is no proof that you are not wintering all right. Same with the bees. They are breathing out moisture all the time, and when it's cold enough you will find that moisture condensing into frost and ice, even though the bees be wintering all right.

Mold in Hives.—Q. What can be done with a colony that has moldy combs when the whole entrance is open? I bored two holes, one on each side of the back part of the cover, about half an inch in diameter, and tacked some screen-wire over the holes, then I placed a telescope cover (of my own make) over it, and packed around it dry moss. Will it work? The bees are in a good shed. They were dying off before I gave them the top ventilation. Now they seem to be doing fine.

A. "The proof of the pudding is the eating of it." If your bees are doing well since you made the change, that is pretty good proof that it is all right. Of course, there must not be too much ventilation, lest the bees be too cold, but ventilation in some form must be sufficient to prevent dampness and mold.

Q. What can I do with moldy comb? Is there any special way to clean comb in which brood and bees have died?

A. Nothing is needed to be done with either moldy combs or those in which bees have died except to give them in care of the bees. They will clean them out in short order. A good way is to put a hive full of such combs under the hive of a strong colony. Then 'let the bees take their time to clean them.

Morning Glory—Q. Does morning glory make nice honey? We have hundreds, yes thousands, of acres here, and the bees seem to work on it some; also carpet-grass. The honey I have extracted is light amber.

A. I have read that it is of fair quality.

Moving Bees.—Q. Will it be safe to move a colony now on a high stand facing the east to a low stand six feet away and facing the southeast? Will many of the bees get lost when they fly out?

A. Something depends upon the weather. If, after moving, the weather is cold enough to confine the bees to the hive for a few days, or if the bees have not been flying for a few days—say a week or so—there will be little trouble about moving bees any distance, great or small. In the particular case you mention there will be no trouble, even if the bees are flying every day, provided no other colony stands within six feet of where the colony in question now stands. Put a slanting board in front of the entrance, so that when they issue they may know that something has been changed in their location. They will then examine their position more carefully before leaving.

Q. Can hives be moved from their original place, say 50 or 100 feet, without confusing the bees? Are there any special rules to be observed?

A. If you move them before they have had a spring flight

there will be no trouble. After that there will be lots of trouble, as the bees returning from the field will return to the old location. One way to help this is to fasten the bees in the hives in the evening, or very early in the morning before moving; after moving, leave them shut in till the middle of the day; pound on the hive until the bees roar, and then let them out, putting in front of each entrance a board for them to knock their noses against.

Q. I shall move my bees about 100 yards this winter, and while I know what the bee-books and journals say about it, I would take it as a personal favor if you would kindly tell how you would manage in moving them. (South Carolina.)

A. Although you do not say so, I suppose you have in mind the question of moving so as to have the least possible loss from bees returning to the old location. If I wanted to move my bees 100 yards, I should wait till winter was nearly over, moving them, as nearly as I could guess, just after they had had their longest imprisonment during the winter, and I would have little fear that any considerable number would return to the old place. The same thing might not work as well with you, as in South Carolina I should expect shorter periods of confinement and more frequent opportunities for flight. I should still work on the same general principle, adding some precaution. Clear up things at the old place, so that if any bee should try to return it would find nothing to look like home. Move the bees in the evening, when all are at home, and fasten them in so none can fly out, but not so as to smother them. Next day, or the first day it is warm enough for them to fly, pound on the hives so as to stir them up thoroughly and set them to roaring. Keep them in suspense for some time, leaving them thus until perhaps noon, if you think there is no danger of smothering, then let them out, and you can expect them to mark their location.

Q. I have six colonies of bees which I keep for pastime and study, as they please me and take up many interesting moments. The hives are scattered, and I would like to have them closer together. One hive is north of my house, three west of the house, about 20 feet from the first; then about 50 feet farther west comes another, and then again about 80 feet west is the last one. The advice I seek is when and how to get these all to the east of the house. I winter them in the cellar. I will greatly appreciate your advice.

A. That's easy. When you put them out in the spring, without any ceremony, you can put them just where you want them. To be sure, some say that bees remember through the winter where their old stands were, but there cannot be much trouble from that, for I have many times put my bees on new stands in the spring without trouble.

Q. As I have to move about three miles March 1st, when do you think would be the best time to move my bees? I have them in boxes with chaff packed around them. Will the bees get ventilation enough from the entrance, or had I better put a screen over the top of the hive in moving them?

A. It will not do to disturb the winter packing any sooner than is necessary, so you had better not move them till bees are flying every few days, possibly in April. If you choose a cool day the entrances will probably give ventilation enough, unless it be that it is less than the equivalent of 3 or 4 square inches. Of course, extra strong colonies may require extra ventilation.

Q. Next spring I want to move 20 colonies in a wagon. When would be the best time to move them, and how would be the best way to load?

A. It doesn't matter such a great deal what time in spring you move them. If you move them when it is freezing hard there is danger that the combs will break. If you move them after they have begun to fly freely you must take the precaution to close the hives the evening before, otherwise you will lose some of the field bees.

Put them in the wagon with the frames running crosswise, as the greatest shaking is from the wagon swinging from side to side.

Q. I would like to move my 15 colonies about 80 to 100 miles from here. I made arrangements to move them in the spring while they would be light and not so crowded, and so that there would still be snow up in the hills to take them on the sleigh where otherwise the road would be rough. The time to go over the snow would take about one day. If I leave the entrance open, also the top, and shut up with screen, put the hives on a spring wagon and some straw under the hives, would this plan work all right?

A. Your plan ought to work all right. There remains the possibility of an unusually warm day occurring during the part of the journey when the bees were on the wagon, making the bees very uneasy. In that case you would quiet them by sprinkling water upon them.

Q. Can bees be moved in the fall, say the last of September or the first of October. I want to move them 75 or 80 miles, either by rail or wagon. Can it be done without damage?

A. For more than 20 years I have moved bees every fall, and never had any trouble. But I moved them only five miles or less. In the fall the combs are heavier with honey than in the spring, and there are also more bees. So you will see that there must be a little more care against breaking combs, as well as a little more care to have plenty of ventilation. Aside from this you ought to have no more trouble in fall than spring. If you can have your choice as to time, it will be well to wait till as much after the first of October as you can, for the cooler it is, the less danger of suffocation, although, of course, if you wait for severe winter weather there would be danger of the combs becoming brittle with the cold, and breaking.

Q. I want to move my bees about 40 miles by waterway to a better location, as the bees are mostly wild and dark. Which is the best way to close the hive and not smother the bees?

A. Use wire-cloth for ventilation. To close the entrance of a hive, take a piece of wire-cloth as long as the inside width of the entrance and 2 or 3 inches wide. Bend it at right angles, and then crowd it into the entrance so it will be wedged fast. But that will not answer if your entrances are like mine, 2 inches deep. In that case take a strip of wire-cloth about 2 inches wider than the depth of your entrance, and as long as the inside width of the entrance. Double over the edge three-quarters of an inch, or an inch, and crease it down flat. Place the wire-cloth against the entrance with the folded edge down at the bottom-board, and nail over the upper part of the wire-cloth a strip of lath with a small nail at each end. If the weather is cool, or if the bees be moved at night, this ventilation at the entrance may be enough. If more is needed, make a frame the same size as the top of the hive, cover it with wire-cloth, and fasten it on top of the hive with screws. If necessary, the cover can be put about 2 inches above this, a block at each corner holding up the cover, being fastened with hive-staples. Even this ventilation, if the weather br hot and the bees kept on the way long, water should be sprayed on them from time to time.

Mustard.—Q. Do you consider mustard a good honey yielder? If so, how does it compare with smartweed in the yield of honey and quality?

A. Mustard is a good honey-plant. Just how it compares in yield and quality with smartweed (by which you probably mean heartsease) could be better told by someone having an equal acre-

age of considerable extent of each. In Europe, rape, which belongs to the same family as mustard, is a honey-plant of great importance. It is possible that mustard would be equally important
if it were cultivated to the same extent. As to quality, Root's "A,
B, C and X, Y, Z of Bee Culture" says: "The honey from these
plants is said to be very light, equal to any in flavor, and to com-
mand the highest price in the market."

Nectar.—Q. The bee gathers nectar from the flowers, which
nectar, after undergoing a chemical process in the bee, becomes
honey. Is not nectar dumped into the comb, then evaporated and
becomes honey?

A. No; if you were to gather nectar and put it in cells it
wouldn't be honey, and if the bee were to dump the nectar into
the cells just the same as it gets it from the flowers it wouldn't
be honey. It must undergo a change in the bee, although that
change may continue afterward.

Noise and Bees.—Q. Would the noise made by moving the
bee-yard bother the bees any? Does noise of any kind bother
bees?

A. Noises in general do not trouble bees, but jarring will irri-
tate them.

Nucleus.—Q. Please explain the meaning of the word nucleus.

A. A nucleus is a baby colony. Just when a nucleus becomes
large enough to be called a colony is not easy to say. Perhaps I
might say it should be called a colony when it has more than
three combs covered with bees; this in summer-time. In the
spring plenty of colonies have only two or three combs covered
with bees.

Q. Would a Danzenbaker hive answer as well as a Langstroth
for a nucleus hive? I use a Danzenbaker altogether. I thought
I would use three frames in each compartment and cut a hole
one-half by two inches in the bottom part of the hive for the middle, as the frames are closed-end frames, and I cannot put the
entrance anywhere else. The outer compartments will be the
same as described in "Fifty Years Among the Bees."

A. Yes, if you are using the Danzenbaker hive, use it for
nuclei also.

Q. Does "baby nucleus" mean simply an ordinary hive with a
few frames, or a small hive full of frames?

A. The term usually applies to a nucleus in a small hive,
with one or two very small frames.

Q. If you had some old nuclei you wished to strengthen by

giving a pound or two of bees purchased from some other bee-keeper, what precautions would you use in uniting the purchased bees to the nuclei?

FIG. 22.—Examining a baby nucleus.

A. Might do one of several ways. One way is to sprinkle all thoroughly with sweetened water. Or, shake them up in a dish-pan till they don't know "where they are at." Or, put the added bees in an upper story for a few days, separated from the lower story by a wire-cloth. During a honey crop they may be given right at the entrance of the colony to be helped.

Q. Is it best to set nuclei quite a distance from strong col-onies?

A. I don't believe it makes much difference.

Q. When forming a single or twin nucleus, which is the better to use, a ripe queen-cell or a virgin queen?

A. There is little to choose. If a cell is given, the young queen is more sure of kind treatment than when a virgin is intro-duced. On the other hand it sometimes happens that the virgin in the cell has imperfect wings, and she may even be dead, and when you give a virgin that has left her cell you know just what you are giving.

Q. Would you advise one-pound packages of bees rather than 1, 2 or 3-frame nuclei?

A The same number of bees will, of course, be worth more with frames of brood than without; but considering the expense of expressage on combs, it is likely that a given amount of money put in bees without combs will be better than the same money put in nuclei.

Q. Is there any trouble with robber-bees bothering new nuclei?

A. I came pretty near saying always. I'll modify that by saying always if honey is not yielding, and care should be taken, even when it is yielding.

Q. Is it possible to winter a 4-frame nucleus packed with chaff in an outside case on the summer stand?

A. It might succeed and it might not. Something depends upon the severity of the winter, and the sheltered location.

Q. Through my carelessness and a poor season I have two weak nuclei at the commencement of cool weather, which I am desirous of wintering over, as they are headed by two of my best queens. How shall I best winter them over? They are of about 3-frame strength. In this locality people winter bees out-of-doors altogether. Our winters are, as a rule, rather open. Sometimes it goes to zero, but that is seldom. The bees have a flight about every two or three weeks.

A. One way is to winter both in the same hive. Put in a division-board that separates the hive in two equal parts, and put the nuclei in these two parts, each nucleus up against the division-board, so that they may have the advantage of the mutual heat from each other. It is possible they might winter through in separate hives, if the hives are well protected. Strengthening each nucleus by giving brood and bees (bees alone if the brood has all hatched out) from other colonies will help their chances if you should try to winter them in separate hives.

Q. Can a nucleus be wintered on top of a strong colony by placing a queen-excluding board between? That is, put two or three 1 or 2-frame nuclei in a hive and put it over a strong colony with plenty of honey?

A. I don't know that anyone has ever tried exactly the thing you mention. Something like it is done in the Alexander method of putting a weak colony over a strong one in the spring; but in that case it is not continued more than three or four weeks. If continued through the winter it is very likely there would be more or less loss of queens. A safer plan would be to use wire-cloth instead of a queen-excluder, so that there would be no communication between the bees below and above. Of course, an entrance to the outside would have to be allowed to each of the nuclei. If outdoors, these entrances should be very small, and the entrance to the lower colony would need be less than with no entrances above.

Observation Hive.—Q. I have an observatory hive in which I expect to put bees this spring. How shall I get them started?

A. There's no trick about it; merely start as you would in any other hive, by putting in the comb of brood with bees and queen. To prevent the bees going back to the old home, set the hive in a dark cellar about four days.

Q. I have Italians and the "blacks," as I call them. I bought an observation hive for one frame and I took out one frame of comb with brood and enough bees to cover the brood well, but the next day the bees came out of the observation hive and into the old hive, and there was not a single bee left. What was the trouble?

A. There is nothing unusual in the case. Take a frame of brood with plenty of bees to cover it, and without any precautions put it in a new hive, and the proper thing on the part of the bees is to go back to their old home. If you had fastened the bees in for about three days they would have staid. If you had taken bees that had been queenless for three days or more, you would have had less trouble. If you had taken the queen with them, returning her after two or three days, more of the bees would have staid. But you probably took them from a hive with a good, strong queen, and they very properly resented such treatment. It is a good plan· to remove the queen after a few days, as it gives you the opportunity of watching the rearing of young queens by the bees.

Q. Do the sides of an observation hive have to be covered with some opaque substance, or will the bees allow the light to penetrate their domicile at all times?

A. It is usual to keep the hive darkened when not under observation, but not absolutely necessary. The bees will daub more propolis on the glass if the light be continuous. They will also worry more.

Q. What is the best location for an observation hive? Would an attic with a northern exposure be best, or what is the best?

A. The best location for an observation hive is one that is the most convenient for the observer, and at the same time comfortable for the bees. The most convenient place for you might be in one of the living rooms, with a bee-opening to the window-sill, and that would likely be comfortable for the bees. But there might be objections to that, such as the meddling of children, driving you to the attic, where there is danger of too great heat on the south side. In some attics the north side would be all right; in others still too hot. To decide just the place for you, conditions, and also premises, must be carefully considered. Keep your observation-hive in some secluded corner, very close to the house, where the bees are not in the way, but where you can watch them at any time.

Q. Do bees winter well in observation-hives?

A. No; although an observation-hive might be constructed, and perhaps some are, so as to be all right for wintering.

A good observation-hive consists of a frame or frames super-posed in which every side, every nook and corner is subject to inspection through the glass. It is, therefore, usually not fitted for proper wintering.

Oilcloth.—Q. Do you think it a good plan to put oilcloth around the hives, leaving an air-space between hive and cloth during hot weather?

A. Some protect the supers in this way to prevent the bees from deserting them during cool nights, but it is doubtful if it would be a good thing. Building paper is better.

Q. My bees are in the cellar, and the oilcloth has not been removed from the frames. Had I better remove it now? I do not know that the bees can get around the ends of the frames, which are 1½ inches from the bottom-board, and the hive is raised one inch.

A. There is danger that moisture will condense upon the oil-cloth and fall in drops upon the cluster of bees. The colder the cellar the more the danger. If you can remove the oilcloth without disturbing the bees much it would be well.

Outapiaries.—Q. Is there any way to judge how many bees to keep at each apiary? Two years ago I had an apiary of 45 colonies; they gathered 22 cases of honey, and they had enough to winter on. A bee-man told me he thought I ought to keep 150 colonies here at this apiary. Would I have gotten as much honey, per colony, as I did with the 45? Then there would be the extra honey for the 150 colonies to live on. I would have more to store honey, but they would have to have honey to live.

A. I have been trying for 50 years, right in the same location, to learn how many colonies could be kept without overstocking. and I don't know yet. One great trouble is that no two years are alike as to yield. In a poor year there may not be nectar enough for 25 colonies, while in the very same spot a good year may give abundance for 100 colonies. And you never can be entirely sure in advance whether the year will be good or bad. If the 22 cases made by your 45 colonies were 22 cases of 24 sections each, that would be about 12 sections per colony. If that was all the surplus that could be stored by decently good bees, it is doubtful that a larger number would have given so very much more. For you are right in counting the honey gathered by the bees for their own use, and it is generally a good deal more than the amount they put in the supers. Suppose it takes 200 pounds per colony for their own use, and that each colony yields a surplus of 100 pounds. Each colony would then gather 300 pounds, which would be 13,500 pounds for 45 colonies. Now suppose the field yields 15,000 pounds. There would be 1,500 pounds that would go to waste, and you might just as well have five colonies more to gather it all. But suppose you plant 150 colonies. They would need 30,000 pounds for their own use. But the field yields only 15,000, and so you would get no surplus and would have to feed 15,000 pounds. Some locations are much better than others. In some parts of Iowa beekeepers harvest big crops with 300 colonies in a single apiary.

Over-stocking.—Q. I have only had my bees about three years. The man I bought them from said he was selling off his bees and was going to Old Mexico, as that was a great bee country. So I bought about one-half of his bees, and he went away and was gone about two years. Then he came back and began to keep bees again. I have four apiaries now. One was doing fairly well, but he has just put a big apiary about one-half mile from mine. We figure on 50 pounds per colony here. Now, what would you eastern beekeepers think of being treated this way? It does not look to me like he or I will get very much honey by having the bees so close together. The locations for bees are about all taken up here, I think. There are some new locations about 18 miles from here. This over-crowding does not look very encouraging to me. What do you think of it? (Arizona.)

A. My thought about it is that this sort of thing makes beekeeping a very uncertain thing to count on. Years ago I took the ground that if ever it was to be a reliable business, a man should have just the same right to his territory as the man who keeps cattle or other live stock. So far as I now remember, not a single man expressed any agreement with me, although since then a good many have. There is quite a general agreement that a man has a prior right morally, although some do not even believe in that. But in matters of business moral rights are not very reliable. I have a moral right to the possession of my horses, but if I had no legal right to them I doubt if I would keep them long. Some day beekeepers may be advanced enough so that a man may be just as safe from interference in his bee-pasture as he now is in his cow-pasture. At present you have no redress, and must just grin and bear it—or else bear it without the grinning.

Paint.—Q. Should I paint my hives all the same color? If so, what color would you advise? I see frome reading the American Bee Journal that some beekeepers advise painting hives different colors, as one color bothers the bees in locating their hive.

A. There would be some advantage to the bees in the way of recognizing their hives if they were of different colors, but it is hardly necessary. Bees locate their hives by means of surrounding objects, and except on a bleak plain utterly without any surrounding objects there is very little difficulty where the hives are 5 feet or more apart from center to center. But you can just as well have double the number of hives on the same ground by having them in pairs. Set two hives close together on the same stand, then leave a space of 2 feet or more, then another pair, and so on. Ground may be still further economized by placing another row close to the first, letting the hives stand back to back.

There is probably no better color, all things considered, than white, using good white lead.

Q. Please advise me relative to the painting of hives with the bees in them, and at what period of the year is it best to do the work? I should also like to know whether or not standard paints are all right to use.

A. You can paint a hive with bees in it at any time when you can paint the outside of a house, and can use any paint proper for the same purpose, with the exception of the part at the entrance where the bees alight. If you put enough drier in the paint used there, and paint in the evening after the bees stop flying, it will be dry enough next morning so the bees will not stick in it.

Q. Should bottom-boards of hives be painted inside?

A. It is not necessary, although, of course, a bottom-board

will last longer if the under side is painted, especially where quite near the ground. But there is no gain in painting the inside.

Painting Hives.—Q. Would you advise me to paint the hives?

A. I don't believe it is best for me; but the majority think better. Painted hives look better and last longer; but I think unpainted are better for the bees.

Q. Why is it that you don't paint hives?

A. Following the teachings of G. M. Doolittle, in whose ideas I have great confidence, I think there is better chance for the moisture to dry out of unpainted 'hives than out of painted ones. I have seen a painted hive in my cellar damp and moldy when all the unpainted ones were in much better condition.

Paralysis (See Bee Paralysis.)

Parcel Post for Honey.—Q. Can extracted honey be sent through the mails in friction-top pails by putting it in wooden boxes, provided the honey is candied solid so that it would not run if the cover was taken off in transit?

A. Yes, such honey can go by parcel post all right.

Parthenogenesis (See Dzierzon Theory.)

Pasturage for Bees (See also Apiary, Location, Outapiary).— Q. Do you plant anything for your bees to work on?' If so, what?

A. After trying many things, I now plant nothing specially for the bees.

Q. Name the kind of pasturage I must have to get good honey. (Illinois.)

A. White clover, basswood, Spanish needle, heartsease, and fruit blossoms are a few of the principal honey-plants in your state.

Pickled-Brood.—.Q Is there a cure for pickled-brood? I know that during a good honey-flow the bees generally get over it. The inspector told me it was pickled-brood. My bees did not get in shape for a crop of honey until about the time it stopped.

A. Pickled brood seems hardly a disease, but is believed by some to be only brood dead through chilling or some other cause, so there is no cure for it, and it needs no cure, disappearing of itself. If you are not sure as to what ails your bees, you had better send a sample of the brood to Dr. E. F. Phillips, Department of Agriculture, Washington, D. C., and after analysis you will be told just what the trouble is. It will cost you nothing, and if you write in advance to Dr. Phillips he will send you a box in which to mail the sample, together with a frank to pay postage.

Pines.—Q. Do bees ever make honey from pines? My bees are bringing in quite a little honey now, August 9, when usually there is nothing in this section except a few cowpeas that about feed the bees. It has been very dry here for three weeks, following an unusually wet spell. During most of this time the majority of the pines in this place have been covered with bees, and a fine-flavored honey is being stored. (Tennessee,)

A. Yes, bees store from pines and, in some parts of Europe, very largely.

Pollen.—Q. My bees are coming in from the fields with their legs loaded with pollen, and there is nothing in bloom here but red elm and a few little wild flowers. Do you think they work on red elm?

A. Yes, bees work on any of the elms. They may also be working on something else that you know nothing about. Bees can beat us humans a long way in finding nectar or pollen.

Q. Last spring my bees gathered pollen from maple trees (not sugar maple) as soon as it was warm enough for them to fly. Will they need to be fed flour?

A. No; with plenty of soft or red maples, they will need no substitute for pollen.

Q. In winter, if bees run out of honey stores, will they feed upon the stored pollen? Is it as good as the honey stores?

A. No; when the honey is all gone they will starve to death, leaving plenty of pollen in the hive.

Q. Where can I get a kind of bee-powder or food that is fed to bees to make them work better and produce more honey? My neighbor uses such, but refuses to tell me where he got it or what it is. It looks something like wheat flour. Bees like it very much.

A. There is no sort of secret powder or food that can be given to bees to make them do more unless it be honey and pollen, and there's no secret about that. The thing probably meant in the present case is some kind of meal used in place of pollen. In the spring, when the weather is good, and yet there is no pollen to be had, set out a box or dish of any kind containing some kind of meal, and the bees will take it in place of pollen. Grain of any kind ground will answer. The kind I have used more than any other is ground oats and corn—the kind that cattle and horses eat, that kind being conveniently on hand. Put a stone or block under one side of the box, and when the bees dig the meal down level, change the stone to the other side. They will dig out all the fine parts, and the coarser parts that are left can be fed to the four-legged stock. But just as soon as they can get the natural pollen they will desert the meal-boxes.

Q. Why is it that in your telling of the use of rye-flour and pea meal for artificial pollen, you never mention wheat flour? Why is wheat flour never mentioned or recommended?

A. It is probably a case of blindly following tradition. My guess would be that wheat is as good as rye; but I never tried either. I know that ground corn and oats do well. Flour is good also.

Q. Will you kindly advise me what to do with extracting-combs that are filled with pollen? Many of mine are so clogged with pollen that I will be compelled to melt them unless there is some way of getting it out.

A. I'm just a bit suspicious that the trouble is not so bad as you suppose, and that if you leave the pollen where it is it will be used up by the bees next year, always supposing it is kept in good condition over winter. It often happens that such pollen is worth more than its weight in honey early in the season. If, however, you want to get the pollen out of the comb some other way than to have the bees eat it out, I'm not sure that I know of any good way. I have known pollen to dry up in the combs so it would shake out.

Q. Will the bees remove pollen from the center combs in the brood-chamber so the queen can have a compact circle to lay in, the combs being filled by queenless bees, caused by the queen being lost in mating? The pollen is fresh, and the cells about half full.

A. Yes, give them time enough and you will find the pollen all out of the middle of the brood-nest, if the queen is prolific.

Q. What can I do to prevent bees from storing pollen in the sections?

A. I know of three things that will encourage pollen and brood in sections. One is to have the brood-chamber too small and crowded. A second is to have very little or no drone-comb in the brood-chamber and small starters in the sections. In that case the bees will build more or less drone-comb in the sections, the queen will go up for the sake of laying in drone-cells, and pollen will follow the brood. A third is to have shallow combs in the brood-chamber. In that case there is danger of pollen in sections even without any brood in them. Evidently, to avoid pollen in sections, we must avoid the three conditions mentioned. I rarely have any trouble in that way, and I use 8-frame hives with frames 9⅛ inches deep, and have the sections filled full with worker-foundation.·

Poplars.—Q. Are poplars good honey-flowers? They are plentiful here.

A. Yes; but the word poplar is used for different trees in different places. What you call poplar in Virginia is probably Liriodendron tulipifera, which is also called tulip tree and white-wood. It is a good honey-tree, although the honey is dark, I think.

Porticos.—Q. Of what use is the portico on some styles of hives?

A. It is supposed to protect from the wind any bees inclined to take a promenade on the alighting-board. It also furnishes a nice protection for spiders, and is not much used nowadays.

Pound Packages.—Q. Could you start a colony with one pound of bees and a queen?

A. Yes; if started early enough in a good season it will make a good colony.

Propolis.—Q. Is there any wax in beeglue or propolis?

A. No, and yes. In pure propolis, of course, there is no wax; but in propolis scraped from sections or frames—indeed as bees use it in general—there is more or less wax, as you will find out if you will melt it.

Q. Should the wax and propolis between the frames be taken off every time the hives are examined?

A. No; you are doing unusually well if you attend to it once a year.

Q. Is there any way of getting the propolis off of fence separators besides the tiresome way of scraping? Will boiling injure the glued joints?

A. I don't know of any better way than to scrape. Boiling in water would dissolve the glue, and would not be a success in removing the propolis.

Q. Is there any sale for propolis, and if so, tell me where I can sell it. I have heard it is worth quite a bit, but never could find out where to sell it, or how much it is worth.

A. I very much doubt if there is any market for propolis. If you have propolis that has been saved from scraping frames, sections, etc., you may find it a paying job to melt the beeswax out of it.

Q. How can I refine propolis and separate it from the wax? Does it lose its aroma when boiled in water?

A. I don't know how to refine propolis. I have separated it from wax by putting it in a dripping-pan in the oven and pouring off the wax; but it doesn't make a perfect job. One would think it would work to boil it in water. Boiling water doesn't seem to hurt the aroma of propolis.

Pumpkin.—Q. Do pumpkin blossoms, nettles, common mint, peppermint, snap-dragon, camomile and love-in-tangle produce honey or pollen for honeybees?

A. Pumpkins and all kindred vines do, also the mints. I don't know about the others.

Punic Bees.—Q. What do you think of Punic bees?

A. The little experience I had with Punics makes me think them hardy and industrious, not with the sweetest of tempers, the worst gluers I ever saw, and capping honey so watery-looking that they are fit for extracted honey only.

Q. Have the Punic bees proved a success in this country, or are they still an experiment?

A. In the experimental stage; reports varying from favorable to extremely unfavorable.

Put-up Plan.—Q. I do not understand the "put-up" plan as per pages 167-8 of "Fifty Years Among the Bees." On page 168 you say: "The cover is put on the supers, and the 'put-up' hive is filled with brood, and is placed over all."

If I were to do this I would put a solid board over the supers. Then I suppose you mean to place a brood-chamber of the "put-up" hive directly over all. But there would not be any place for the bees to get out.

Should a queen-excluding board be over the supers? Or, if a solid board, would I have to arrange the brood-chamber on top so as to leave an entrance for the bees to get out?

A. You have it straight until you say, "I suppose you mean to put the brood-chamber of the 'put-up' hive over all." I think the whole thing will become clear if you note that I do not say "brood-chamber," but that the hive is placed over all, and then remember that ordinarily when we talk about a hive we mean not merely the brood-chamber or hive-body, but the bottom-board along with it.

To be specific about it, the lower hive has placed on it the super, or supers, and these are covered up just as they would be if no other hive was to be placed over. Then on top of this is placed the put-up hive with its bottom-board and its cover. This, you will see, leaves an entrance for the bees in the upper hive, just as there would be if, instead of being put up, it were set on a stand down on the ground. There is no possible communication between the two hives, and if a bee goes from one to the other it can only do so by going out at one entrance and going in at the other.

Q. Would it not answer to put up the queen as soon as queen cells with larvæ in them are seen, instead of waiting and watching for swarms?

A. Yes, it works well, although I have not had as much experience with that plan as with waiting for the bees to swarm.

Q. When you put down the queen again, is there no danger of her being balled?

A. I do not recall that she was ever balled, to my knowledge.

Queens.—Q. How can you tell a queen from the rest of the bees?

A. Look for a bee longer than the rest, and with wings that look too short for the length of its abdomen. You'll not be likely to miss it the first time you see it.

Q. What is meant by fertile queen, and virgin queen?

A. A queen that has met a drone, a normal laying queen, is a fertile queen. A virgin queen is one which has not yet been mated.

Q. In the advertisements of queen-breeders, the following terms are used, which I do not clearly understand: Tested, Untested, Select Tested, Select Untested Queens, and Breeders.

A. A tested queen is one whose progeny shows she has mated with a drone of her own race. In the case of an Italian queen you will see that that will mean that the worker progeny of the young queen shows three yellow bands.

An untested queen is usually one sold as soon as she begins to lay, and so nothing is yet known as to the appearance of her progeny. An untested queen, of course, can be sold at a less price than a tested one, and that for two reasons. In the first place, it saves the expense of keeping the queen some three weeks, if she is sold untested. In the second place, if queens are kept until tested, those which do not come up to the test must be rejected or sold at a lower price as mismated, while all will be sold at the same price if sold while untested.

A select queen, either tested or untested, is one that is selected because she is unusually good in appearance. However it may be with a select tested queen, a select untested queen has nothing but her looks to entitle her to a higher price, for nothing can yet be told about the looks of her unborn progeny, to say nothing about the performance of the same.

A breeder is one that is considerably better than the average, and so, of unusual value to breed from.

You will see that there is chance for a good deal of looseness in the whole business, especially as good looks and good behavior do not always go together. "Handsome is that handsome does."

Q. Is there any way to tell how good the queen is in a weak colony, during brood-rearing?

A. No. She may lay enough eggs to keep a weak colony supplied, but not enough for a strong colony. Yet even in a weak colony a very poor queen may not keep the cells filled with eggs in an orderly manner, but will skip more or less cells. Even in a strong colony you cannot tell how good a queen is merely by looking at her brood. The most prolific queen is not by any means always the best. To learn how good a queen is you must wait to see how much honey her bees will store compared with others.

Q. I put a full depth super on top of one of my colonies, and an examination afterwards showed the queen was rearing brood very extensively in the upper story, and later on I examined the lower story and found that she had deserted it altogether, and the cells were all full of pollen. Could you tell how this could be avoided?

A. A queen-excluder of perforated zinc will prevent the queen from going up.

Q. Will strange queens sometimes unite with a queenless colony?

A. Yes, sometimes it happens that a young queen may go into another hive than her own.

Q. One morning I found six dead queens in front of the hives. Why do they have so many queens?

A. Nature generally makes bountiful provision against danger of failure. Take an apple tree, one that is thoroughly filled with blossoms. What if every blossom should produce an apple? If there's one apple for every ten blossoms there will be a heavy crop. But if there should be merely enough blossoms for each expected apple, something might happen to a good many of them, and there would be a shortage in the crop. Same way with the bees. Hundreds of drones are reared for every one needed, so there shall be no lack, and a number of extra young queens are also reared. At the last there may be a duel to settle which one of these young queens shall reign, and that gives you a chance to have the most vigorous one left.

Q. Can you give me the cause for a young Italian queen hatching with only a part of a wing?

A. Insufficient nourishment or a slight chilling, which may occur in a weak colony. Even in a strong colony a cell on the lower edge of the comb might be chilled on a very cold night. It has been said that letting a queen-cell fall, or shaking it might result in crippled legs or wings. In rare cases, also, a moth-worm may have traveled through the wall of the cell and clipped the queen's wing.

Queens, Age of.—Q. How long is the life of the average queen?

A. Perhaps about two years, varying from a few weeks to four or five years.

Q. Is there any way to tell the age of a queen, and also how old should a queen be allowed to get?

A. There is no certain way to tell by the looks of a queen how old she is. After you have some experience you will be able to make a fair guess as to whether a queen is old or young, as an old queen is more inclined to have a shiny look, because her plumage is worn away. Sometimes, however, a young queen has the same look. An old queen is not likely to move about on the combs in as lively a manner as a young one.

There are different views as to how old a queen should be allowed to become. Some think not more than two years. In my own practice I allow her to live as long as she will, for when she gets too old the bees will supersede her without any interference on my part. Of course, if she is unsatisfactory in any way, I get rid of her as soon as I can.

Queen Balled.—Q. What is meant by the bees "balling a queen?"

A. Very much what the word indicates; hostile bees will grab hold of the queen at different parts until there are bees all about her; then other bees will seize those that have hold of the queen, until there is a ball of them as large as a hickory-nut or larger.

Q. What should be done to a queen if balled?

A. Throw the ball in a dish of water and the bees will leave her. Or, you may smoke the ball; but hold the smoker at a distance, for if hot smoke is thrown on the ball the bees will sting her.

Q. What should I have done when I saw the bees balling the queen, other than I did—pour a little warm syrup on them and close the hive? Honey was coming in fairly well, and I did not use smoke—just a whiff over the tops of the frames.

A. You did the right thing. When bees ball their own queen, if the hive be quickly and quietly closed, rarely does any trouble follow. If you want to be so careful as to guard against the rare case that sometimes happens, you can cage the queen and let the bees liberate her by eating out a plug of candy.

Q. Last summer I had six swarms come out and go away to-gether (unclipped queens), and some of the queens were balled and killed. What can one do to separate them?

A. You can pick out each ball, put it in a hive, and then distribute to each ball its proportion of bees; for a queen is not likely to be injured in a ball until you have time to make the dis-tribution.

Q. I lost several queens last season when they returned from their wedding flight. The bees balled them. I have found them balled when they were not over a week old. When I took some bees and a virgin queen and made a nucleus, the first queen would be mated all right. It was the second queen that got killed. Can you tell me why the bees killed the queens?

A. It is said that the bees attack the queen because she has acquired a strange scent. But there may be some question whether in returning from her wedding-flight she is likely to be killed by her own bees if the beekeeper himself does not meddle. Bees sometimes ball their old laying queen, and when I have found them doing so, I have always made it a rule to close the hive as quickly and quietly as possible, leaving the bees entirely to their own devices, and on looking in the hive a few days later everything would be found all right. If you try to rescue the queen from the balling bees, you stand a pretty good chance of having her killed. Why may it not be the same way when bees ball a queen that has just mated?

Queens, Buying.—Q. Where can I get a first-class Italian queen, free of disease?

A. In the proper season there are always found in the Ameri-can Bee Journal, advertisements of those who have queens for sale, and these may be relied upon as free from disease. A man who would send out a queen from diseased stock would steal.

Q. How soon in the spring are queen-breeders ready for mailing tested queens?

A. In the South I think they can ship in March tested queens of the previous year.

Q. How is it that most of the queen-breeders advertise queens for sale and none can supply the beekeeper with queens early, but only want their orders booked early, and maybe have the queens forwarded the latter part of May or middle of June, the time when every beekeeper has plenty of queen material to supply himself?

A. Don't be too hard on the queen-breeders; you may sometime be one yourself. It is all right to book orders to be filled as fast as possible, provided it is an understood thing that they are to be so filled. If, however, he advertises to send queens by return mail, and then delays, he's not giving you a square deal. It looks a little as if your idea was that when you order a queen you should always get it by return mail. It would be difficult for a man to treat all of his customers in that way. He would be obliged to have a stock of queens on hand before he made such an agreement; he would have no way of knowing how many to have in advance, and might be overstocked at a loss. You can, however, say when ordering, "If you cannot send a queen at such a time, return money," and then there could be no complaint on either side.

You say they send queens when every beekeeper has plenty of material to supply himself. Pray tell me how a queen-breeder can have material earlier than the beekeeper. You and I can have material as early as any, and can rear queens as early; but we may want to buy queens for other reasons. Moreover, I wouldn't give 30 cents a dozen for queens reared too early, no matter who rears them.

Queens, Caging.—Q. How long is it safe to keep a queen caged? Must she have attendants as in shipping?

A. It is generally neither necessary nor desirable to have her caged more than ten days; but I have known a queen to be caged double that time in her hive without appearing to be hurt by it. No need of any attendants in the cage; they are likely to die in the cage and thus be a damage.

Q. How long may I keep queens caged (after they have commenced to lay) without danger of injuring them?

A. No doubt something depends upon circumstances. If a queen should be caged in a hive among her own bees, so that they can feed her, she would likely endure confinement several times as long as she would if the cage were left out of the hive with candy for the queen to eat. I have often had a queen caged in her hive ten days or so with no apparent harm, and my guess would be that she might stand it three to five times as long. Caged outside the hive, ten days might be all or more than she would stand.

Q. Can you cage a queen and put her in a colony having a laying queen? If so, how long can she be kept there?

A. Yes, and she may remain weeks, or she may be dead in a few days. She will be more sure to remain in safety if the cage is provisioned than if she has to depend upon the bees to feed her.

Q. Wouldn't it be better to cage queens on a comb of unsealed honey, on the push-in-cage method, than in cages with candy?

A. The way you suggest would be better if the queen be caged in a strange colony; if caged among her own bees there would be no advantage in it. For in that case the bees feed the queen, which is probably better than for her to feed herself.

Q. What is the proper procedure necessary in the caging of a queen over another hive, as in the case of taking one out for ten days or so, in the several different methods of management? What kind of a cage is used, and how and what is the queen fed? Is the common Benton mailing cage all right when provided with good candy?

A. Any cage that will go easily between the combs will answer, such as the Miller cage. The Benton cage is too bulky. Sometimes, however, instead of being put between the combs, the cage is merely thrust into the entrance of the hive. No need of any food in the cage; the bees will feed the queen.

Q. In sending queens by mail, what are escort bees put in for, to keep up the temperature, or feed the queen?

A. The escort bees feed the queen and keep up the temperature, and it is quite possible that they serve an important purpose in keeping up her spirits by their genial company.

Queens, Clipping.—Q. Is there any danger of clipping the queen's wing too soon or before she takes her mating flight?

A. Great danger. If you clip her before mating she will be a drone-layer, if she lays at all.

Q. What is the advantage of clipping the queen's wings?

A. The advantage is that a prime swarm with a clipped queen will return to its hive because the queen cannot go with it.

Q. How do you clip a queen's wings? Is it good policy to do so?

A. Probably the majority think it is good policy to clip. Mr. Doolittle catches a queen by one wing, lets her hold to the comb with her feet, and with a very sharp knife cuts the wing against thumb and finger. Probably a larger number, myself in the number, use a pair of scissors, holding the queen by the thorax (not by the abdomen or hinder part) between the thumb and finger of the left hand, and cutting off most of the two wings on one side.

Q. If a clipped queen swarmed from a hive upon a high stand and fell to the ground in the absence of the apiarist and could not get back, would the swarm return to the old hive, and would they, on finding their queen absent, proceed to rear a new queen in her place, or what would happen?

A. The swarm would return to the hive, in which there are already a number of young queens in their cells. The first of these will emerge from its cell in a little more than a week, generally, and a swarm is likely to issue with her.

Q. Clipping queen's wings, as I have repeatedly read in your journal, is in vogue among American beekeepers. I would like to make a trial of it in the spring, but have some misgivings. Can one be sure that the issuing swarm will find and cluster about the queen, which, perhaps, has fallen upon the ground a few steps from the bee-house? Or can it also happen that the swarm does not find the queen, and consequently returns to the hive from which it issued? (Germany.)

A. When the swarm issues, of course the clipped queen falls on the ground. If there is no one on hand to pick up the queen it very rarely happens that the swarm finds her and clusters about her. Indeed, in all my experience I never knew such a case. Sometimes the queen will be found at a little distance with a little cluster about her, perhaps as big as a walnut. Generally, however, she will be entirely alone. The swarm will return to the hive, perhaps in less than five minutes, after circling around in the air for a little time, and will pay no attention to the queen, even if she be quite near the hive on the ground, its only desire, apparently, being to hurry back into the hive as soon as possible. Often the swarm will cluster on a tree, just the same as if the queen were along, and it may remain clustered there 5, 10, 15 minutes or longer. In most cases the queen will find her way back into the hive if she is left to herself. The business of the beekeeper, however, is to pick her up, put her in a cage, move the old hive away, and put an empty one in its place, and then, when the swarm returns, to let the queen run into the hive with the swarm.

Q. If the queen's wings are clipped and queen-cells are cut out every ten days, will that prevent swarming?

A. The clipping of the queen's wings will not make a particle of difference about a swarm issuing. A swarm will issue exactly the same as if the wings were whole. Cutting out queen-cells every ten days may make a great deal of difference and it may make a very little. In the ordinary course of events a prime swarm is likely to issue when the first queen-cell is sealed. If, at any time before this, you cut out all cells that are started, the bees will be likely to start fresh cells, but this second time they may not wait for the sealing of cells, and the oftener you cut out cells the more eager they may be to swarm, so that finally a swarm may issue immediately after you have cut out cells. Sometimes, however, cutting out cells once or twice in the season may prevent swarming entirely. I think the character of the bees has something to do in the case. Some bees are more given to swarming than others.

Q. Will a queen's wings grow again after they are clipped?

A. A queen's wing that is clipped will not grow again; never, never; no, not the least little bit.

Queens, Chilled.—Q. Will a queen that was chilled coming through the mails be all right next spring?

A. Hardly; but if you want to breed from her you may get good stock, even if she lays so poorly as to be of little value for honey.

Queens, Color of.—Q. Do queens change color or get much larger?

A. There is considerable change in the appearance of a queen. After she is three or four days old she is smaller than when she first leaves the cell, and will be larger after she gets to laying. A queen often is darker after having gone through the mails.

Q. I am told that "the color of a queen has nothing to do with the bees she will rear;" that "pure Italian queens may be yellow, leather-colored, or jet black, but their bees will be yellow." Is this so?

A. That's not so very far from the truth. Some of the best Italian queens are quite dark, although their workers are yellow.

Queens Destroying Cells.—Q. Will you explain what is to me still a contradictory mysticism? (a) It is said that the first queen out destroys the other queens before they emerge; hence, there should not be afterswarming.

(b) Yet the very fact of there being afterswarming shows that the first queen does not stay to destroy subsequent ones, but one flies off after the other.

A. There is nothing mystical nor difficult of understanding when you get the whole story. When a virgin emerges from her cell, her first care is to find the cells of her younger royal sisters, with full intent to murder them in their cradles. With such frenzy does she seem possessed in this regard that I have many a time seen it the case that when a sealed cell was caged, the virgin, after gnawing her way out would dig a hole in the side of the empty cell, just as she would if a live virgin were in it. Always you may count on this murderous impulse on the part of this royal young personage, and if she were left to have her own way there would never be any afterswarming.

Now, however, comes the part that you have left out. She does not always have her own way. In fact, calling her a "queen" is a neat little fiction; the term "slave" would be about as appropriate. The government in the hive is not a monarchy, but a democracy of the most democratic sort, run by a lot of suffragists, and the male person has no vote. If the workers vote that the time has not yet come for the destruction of the young rivals, then a committee stands guard over each cell, driving away the young queen as often as she makes an attack.

In the meantime several of the occupants of the cells may become sufficiently matured to emerge, but they are not allowed to do so. The guards maintain a neutrality strict enough to suit President Wilson; they will not let the young queen get out of the cell, although she may have the capping of her cell gnawed away all but a slight hinge; and no more will they allow the queen at liberty to get at the defenseless sisters in their cells. The free queen runs about frantically from one cell to another, at intervals crying, "Pe-e-e-ep, pe-e-ep, pe-ep, peep," in a shrill voice, each shorter than the preceding one, and then the prisoners reply in a coarser tone, an apparently hurried "Quahk, quahk, quahk," and this piping and quahking will be kept up until a swarm emerges with the free queen. Then it depends upon the vote of the suffragists what further shall be done. If they vote for further swarming a single virgin is allowed to emerge from her cell, and she in turn will go through the same performance as the one who preceded her. But if the vote is for no further swarming, then the guards relax, allowing the cells to be attacked, and also allowing their inmates to emerge. Then there will be a free-for-all fight, one after the other each queen will be killed until only one is left, the victor in each case coming off unscathed. Sometimes a number of the virgins will go off with the swarm, where they can settle their differences as well as if they had stayed in the old home.

Queen, Failing.—Q. Please state some of the indications of a poor, failing, or old queen.

A. Some of the brood in worker-cells may be drone-brood, as shown by the raised cappings of the cells; the brood may be scattering, or it may be scanty.

Queens, Feeding.—Q. Can a queen eat as other bees or do the bees have to feed her? It is said that the bees feed the queen.

A. A queen can eat as other bees, as you can easily determine by caging one for a short time and then offering her a little honey. During the time of year when she is not laying she may help herself like other bees, but in the season of busy laying the bees feed her with food that is no longer undigested. If she were obliged to digest all the food she takes during heavy laying, I'm afraid the daily quota of eggs would decline very suddenly.

Queen, Finding.—Q. Can you give any suggestions to a novice as to how to find the queen?

A. Experience is the best thing. After some practice you'll spot a queen on a comb very readily. Don't do anything to set the bees to running. If they get to running, you may as well close the hive till another time. The two things most likely to set them to running are too much smoke and too rough handling. So use just as little smoke as will keep the bees under subjection, and be slow and gentle in all your movements. G. M. Doolittle says that from 9 o'clock till 3 the queen is most likely to be found on the outside of the comb that has brood in, either on one side or the other. If you lift out two or three frames and set them in an empty hive, that gives you room in the hive to glance over one side of each comb before you touch it at all. That is, when you lift out a frame, before carefully looking it over, glance over the exposed side of the next frame in the hive. Often you may see the queen thus in the hive, when with gentle haste you will put down the frame in your hand and lift out the one with the queen. After looking over the combs two or three times without finding the queen, it is generally as well to close the hive till an hour or two later, or till another day.

Q. I sent for a queen the past summer. I took all the frames out of the hive, but I could not find the old queen, so I put the new queen in and they killed her. How can I use a queen-trap to find her?

A. Fasten your trap at the entrance of the old hive. Lift out all the frames with adhering bees, and put them in an empty body close by. After all the frames are taken out of the hive, make sure that the queen is not in it, if necessary brushing all the remaining bees out of the hive upon the combs. Now lift one of the combs, shake and brush all the bees from it upon the ground in front of the old hive, and as soon as you have all the bees off the comb put the comb in the old hive. Proceed with all the combs in the same way, brushing all the bees in front of the trapped entrance, and putting the brushed combs into the old hive. The bees will crawl into the hive through the trap, and Madam Queen will be found trying to get in the same place.

Q. What is the best way to find and catch the queen in a box-hive?

A. Drum the bees out, put them in a movable-frame hive with a frame of brood in it, give the bees time to settle, and then look on the frame of brood for the queen. Or, you may sift out the queen with an excluder. Usually if you shake the bees in front of the hive you will see the queen running toward the entrance.

Queen Introduction.—Q. Can I introduce new queens at any time, and how will I have to proceed?

A. Yes, you can introduce queens any time, so long as the weather is warm. The proceeding is the same as at any other time, but introducing is not always so successful in a dearth as when honey is coming in freely.

Q. Could I safely introduce a new queen to a swarm hanging to a limb, by killing their queen and placing the new queen on the cluster of bees?

A. It might succeed, sometimes.

Q. Is it any safer to introduce queens to one or two-frame nuclei in place of a four-frame or a colony of bees?

A. Perhaps there is no difference, but if there is any difference it is safer to introduce to the weaker.

Q. What difference, if any, is there in acceptance of a queen in a colony that has been queenless for some time (no laying worker), and in case of increase by division of a colony, as to queen given to the queenless part?

A. Introduction would be quite a bit more likely to be successful in the second than the first case. It is generally found that it is more difficult to introduce a queen to a colony that has been queenless for some time than to one from which the queen has been recently removed. The reason may be because of the age of the bees, for it is the older bees that make trouble when a new ruler is introduced. It is better that they should not be queenless for any length of time.

Q. How should bees act when favorable to accepting a queen introduced to a colony that has been queenless perhaps ever since the swarm was hived last May? (October.)

A. It is easier to tell by looking at them whether they feel like accepting her than it is to tell how one tells. If the bees are hostile to her, they may be grasping the wires of the cage as if trying hard to get at the queen, while if they feel kindly toward her they will sit quietly and loosely on the cage. That's not telling you very much, is it? Well, I may as well tell you that if the case were right before me, I couldn't always tell for certain. They might appear to be looking as sweet as you please at the queen, with murder in their hearts all the while.

Q. I have had trouble in introducing laying queens on account of the bees starting cells. I have always lost about half of the queens I tried to introduce. Would it be perfectly safe to shake about a pound of bees taken from three or four different colonies into an empty hive containing about three combs with no brood, and confine these bees three or four days; in the meantime introduce a queen in the regular way, brood to be given later? Would these bees be likely to swarm out after they were released?

A. The plan has been used, the bees being put in the cellar or other dark place. They ought not to swarm out afterward.

Let me give you one of the kinks I have used in introducing a valuable queen. It is the old bees, and not the youngsters, that make trouble when a step-mother is given them. So the thing to do is to get the field-bees out of the hive before the queen is given. That is a thing very easily done. Just set the hive in a new place, and leave on the old stand a hive with one of the brood-combs. When the gatherers return from the field they will go to the old stand, and in 24 hours the old hive will have in it no bees more than 16 days old. As a matter of convenience, I lift the old hive from its stand, setting it close by; put the new hive with one frame of brood on the old stand, put on this the cover, and then set the old hive over all. At the same time the caged queen is put into the upper hive. By the time the bees have eaten out the candy and liberated the queen, or some time before it, all the field-bees have joined the lower hive, and the queen is kindly .received by the younger bees. In two or three days, when the queen has begun to lay, the hive may be returned to its original place, and the fielders will make no trouble when they enter.

Q. What is the best way to introduce a valuable queen?

A. With a very valuable queen, if you want to be entirely safe, proceed in this way: Put two, three, or more frames of brood in an upper story over a strong colony, having a queen-excluder between the two stories. In about eight days all the brood will be sealed. Now lift the upper story, take away the excluder, and cover the hive with wire-cloth, which will not admit the passage of a bee. Over the wire-cloth set an empty hive-body. One by one lift the frames out of the removed upper story, brushing off upon the ground in front of the hive all the bees from each comb, and putting the brushed combs into the empty upper story. Put your new queen into the upper story and cover up, making very sure that not a bee can get in or out. Your queen is now alone in the upper story, but will probably have company within five minutes, for young bees will be hatching out constantly from the sealed brood. No bee can get from one story to the other, but the heat can rise from below to keep the upper story warm. In about five days you can set this upper story on a new stand, giving it entrance for only one bee at a time. If your bees act like mine have done, and the circumstances are favorable, before night you will see some of the five-day-old bees entering the hive with pollen on their legs.

Q. Please explain the Abbott plan of introducing queens.

A. Put the new queen in a hive with a provisioned cage with the candy protected so the bees of the hive cannot get at it. In about two days remove the old queen and give the bees access to the candy so they may liberate the queen.

Q. I wish you would explain as clearly as possible how to introduce a queen by the smoke method.

A. In "Gleanings in Bee Culture," what you call the smoke plan of introduction was thus given by Arthur C. Miller, of Rhode Island: "A colony to receive a queen has the entrance reduced to about a square inch with whatever is convenient, as grass, weeds, rags, or wood, and then about three puffs of thick, white smoke—because such smoke is safe—is blown in and the entrance closed. It should be explained that there is a seven-eighths-inch space below the frames, so that the smoke blown in at the entrance readily spreads and penetrates to all parts of the hive. In from 15 to 20 seconds that colony will be roaring. The small space at the entrance is now opened; the queen is run in, followed by a gentle puff of smoke, and the space again closed and left closed for about ten minutes, when it is reopened and the bees are allowed to ventilate and to quiet down. The full entrance is not given for an hour or more or even until next day."

Q. What is the Sibbald quick method of introducing queens?

A. Hunt the queen out that is to be removed and put her in a wire cage on top of the frames. Then the queen that is to be introduced is laid on top of the same frames, too, and left till evening. Now remove the old queen and put the new queen in the cage from which the old queen has just been taken, and over the end of the opening fasten a piece of comb foundation. Place on the frames again, after punching a few small holes with a pin through the foundation and let the bees release the queen. Sometimes Mr. Sibbald rubs the dead body of the old queen, that has just been killed, over the outside of the cage she has just come out of.

Q. How soon after introducing a queen is it safe to open the hive to see if she is all right?

A. It is a little safer not to disturb the colony for three or four days.

Queens, Keeping.—Q. The bees will take care of their own queen in a cage. But if she is caged and put in another colony above the excluder, will those strange bees take care of her?

A. Generally there will be some bees so good-natured as to feed a strange queen; but it is safer to have the cage provisioned, and then the queen can feed herself.

Q. I have always been puzzled how to keep a lot of queens when not having immediate use for them. You stated once about the maximum length of time one could keep queens in cages without danger to their laying powers. I suppose while so caged they do not lay any eggs. But even the interruption in laying while queens are in the mails is said to be harmful.

A. In the case you speak of, the queens were kept in small cages in a small colony. This was in the spring when there was no heavy laying yet, and I doubt if the queens were at all injured by being kept from laying. My guess would be that a queen, or a number of queens, might be thus kept safely for a month, perhaps two months, in a queenless colony, or a queen-right colony, if the bees would feed her. Indeed, she might be kept in a candied cage if the bees did not feed her, only in that case bees having a queen of their own might be hostile to her, and this nervous irritation might be bad for a queen. I am not sure that it has ever been claimed that the cessation from laying was an injury to queens sent through the mails. It doesn't hurt

FIG. 23.—The queen laying, surrounded by her retinue of bees.

a queen to remain all winter without laying. Nor is it likely she is injured by ceasing to lay in a dearth long continued. She may be injured by being jarred and frightened in the mails, by sudden cessation from laying, and especially by being flung about when heavy with eggs.

Queens, Laying.—Q. How soon will a queen begin to lay after being fertilized?

A. Generally in two or three days, but she may be longer.

Q. I have an observation hive. The bees were put into this hive about June 1, and I have been looking closely for the queen, but have never seen her. Is she covered by the workers while laying. They have brood and honey sealed.

A. No, she is not covered when laying; but she may be hidden under a mass of bees when not laying. It is a little strange that you have not seen her; but if eggs are present she must be there, and if you persevere you will probably see her.

Q. What kind of a queenbee is it that lays part drone and part worker-eggs in worker-cells; is she an old queen, young, or not fertilized? I bought 50 queens this spring, from one of the most popular queen-breeders in the South, and three of them lay part drone and part worker-eggs in worker-cells, and one more laid all drone-eggs in worker-cells.

A. An old, played-out queen may begin laying occasionally a drone-egg in a worker-cell, and gradually increase until she lays nothing but drone-eggs. But this is by no means always the case with old queens. Occasionally a young queen begins laying without being fertilized, and, of course, will lay only drone-eggs. Sometimes a young queen lays part drone-eggs in worker-cells, either because imperfectly fertilized or on account of some functional disability. Sometimes a young queen lays drone-eggs for a while, and then lays worker-eggs all right.

You do not say what kind of queens you bought, but buying as many as 50 at a time it is practically certain that you bought them as untested queens. That would rule out the chance of their being old queens, always supposing you bought from an honest man. An untested queen is generally shipped as soon as convenient after she begins to lay, and all that the breeder is supposed to know about her is that she is reared from a good mother, that she is physically perfect so far as appearances go, and that she has begun to lay. The purchaser takes his chances on whether she is purely mated or whether the eggs she lays in worker-cells will all produce worker-bees, unless, indeed, they are sold as warranted queens. Yet it is probably not often that so many as 4 out of 50 turn out badly.

Queens Leaving Hive.—Q. While trapping drones this spring, I caught a queen in the trap. Does a queen ever leave the hive except with a swarm?

A. She leaves the hive also on her wedding trip.

Queens, Mating.—Q. Will a queen mate with a drone if she is never allowed to leave the place where she is confined with a drone?

A. If you mean confined to the hive, no. It is possible she might mate if confined in a tent, but it would have to be an immense tent.

Q. Do queens always mate with a drone in the air?

A. Yes.

Q. If a queen is never allowed to mate with a drone, would she lay fertile eggs?

A. If she lays eggs at all, they will produce only drones.

Q. How many times does a queen mate?

A. Once for life; but some cases have been reported in which a queen mated the second time. She may, however, make several flights before mating.

Q. How long will it take after a queen is hatched for her to mate?

A. Five days or longer.

Q. Do you agree that a queen is never mated after she is two or three weeks old? Last March I had a colony of bees supersede its queen, and this colony contained just a small patch of drone-brood which did not hatch till the queen was about 10 days old, and there was no other drone-brood in the yard. The queen commenced to lay when she was about two months old, and now she is the mother of one of the strongest colonies. I give this simply for what it is worth. I examined this colony once every two days, till the queen started to lay, and so these figures are accurate.

A. As a general rule a queen is never mated after she is 10 days old—perhaps not after she is a week old. But there are exceptions, and how far those exceptions extend I don't know. If your queen did not lay till two months old, she may have been fertilized only three days before she began to lay, and she may have been fertilized sooner, but likely she was at least a month old when fertilized. This is exceptional.

Q. Do you think queens would mate with drones a mile away? There is a big woods between us.

A. Yes, a distance greater than that would not prevent mating.

Q. How far away from other bees would we have to place a colony to insure pure mating?

A. You might be safe at two miles, but to be entirely safe you might have to be five miles or more. No one knows exactly how far.

Q. Please say how queen-breeders mate queens purely while bees of other "nationalities" are present.

A. They don't; at least not always. For if it is desired to keep a certain kind pure, they do not have any other kind in the apiary. But something may be done toward getting what you want in this way:

Put in the cellar the hives containing the drones and the young queens. After it is too late in the day for other drones to fly, take out the cellared hives, and incite them to fly by feeding. You may be a little more sure of this if the cellaring has continued two or three days. You may also succeed by taking them out in the morning, so as to get them to fly before other drones are out.

Q. I shall want to clip the queen's wing when she becomes fertile. When should this be done?

A. She is likely to begin laying when about ten days old, although it may be a day or two less, and it may be several days more. Do not clip her till you are sure that she is laying regularly in the combs.

Queen Nursery.—Q. I want to ask about the Stanley nursery for queens. I have Dadant, Hutchinson and Root on bees, but none of them has anything about it. I would like to know where one can be procured or how made. I am anxious to have one.

A. The essential part of a Stanley queen-nursery looks like a cartridge shell for a gun. The shell is made of excluder-zinc, and is 2 inches long, with an inside diameter of three-quarters of an inch. The perforations of the zinc run transversely. It is simply a piece of excluder-zinc 2.36 inches long and 2 inches wide, rolled up into cylindrical form and soldered together. The two ends are closed by common gun-wads. The workers have free entrance to the cylinders, while no queen can enter to make an attack. In the little experience I had with them I found that the young queens were sometimes killed by getting caught in the slots, but not often. They have, on the other hand, the advantage over other nurseries that the workers can have free access to the cells, and it is claimed, especially in Europe, that the close contact of the workers has a very important influence on the occupants of the cells. A number of these cartridges—I think 48—may be contained in an ordinary Langstroth brood-frame, and be put between the brood-combs in a hive. They can probably be had from the inventor.

Queens, Northern and Southern Bees.—Q. Would it do to send to Texas, or other warm countries, for queens? Would they stand the cold up here in New York state and be hardy?

A. So far as I know, queens from the South do just as well as those reared farther North, and are just as hardy.

It is well known that, in general, each region has plants and animals adapted to its particular climate and locality, and those of tropical regions do not well endure the rigors of the far North. So it is natural to suppose that bees in the South become less hardy. But characteristics do not change over night, and if bees become less hardy in the South it would be only through a long course of years. Even if a southern breeder should have stock that had been bred in the South for a hundred years, if there was any suspicion that it had become less hardy, it would be the work of a few weeks at most to have that stock entirely changed through getting one or more queens from the North.

So the usual reply that queens reared in the South are just as hardy as those reared in the North may be counted correct for all practical purposes. Italian bees are from a country with a warm climate. It freezes but little in any part of Italy, and the climate is certainly less severe than that of Texas.

Queen-Rearing.—Q. How late in the season can queens be reared and mated?

A. That depends on the season. If honey is yielding, any time through September. But you are not likely to have good queens if you rear them too late, and losses on wedding flights will be greater.

Q. I have been considerably puzzled by a case called to my attention in which a party claims that a hive of bees swarmed with a virgin queen, leaving a clipped queen at the head of the colony. I have been under the impression that the bees or the virgin queen generally killed the old queen on account of her inability to leave with the swarm.

A. You are right in your impression as to the bees or the virgin putting out of the way the old queen; at any rate, when a colony with a clipped queen swarms, and the beekeeper does not interfere, you may count upon the old queen turning up missing a week or more after the issuing of the prime swarm, and the colony swarming with the virgin. But I think I have seen reports of rare exceptions. At any rate, it is not impossible that the old queen might be suffered to remain, perhaps both queen and virgin going with the swarm, and then the old queen crawling back into the hive.

Q. My frames are about 9 inches square, inside measure. I have some small hives that hold four and five frames each. Will they rear strong queens if given eggs? These hives are used to build up.

A. A hive containing four or five frames, each 9 inches square, would not hold a very strong colony, and a queen reared in it would not be so good as one reared in a strong colony, at least up to the time of sealing the cell. After the queen-cell is sealed it is not so important that the cell be in a strong colony and in hot weather it will do very well to be in a nucleus.

Q. Do you consider forced queen-rearing (as used by those who transfer the larvæ) as good as natural methods given by you in your book? Are the queens as long-lived and as prolific?

A. In the hands of skillful men I don't see why just as good queens cannot be reared by the methods in vogue among queen-breeders but I don't see how they can be any better. But I would lay stress upon having cells started under favorable circumstances, with a good yield of honey, and in a colony in the humor for starting cells. No colony is too strong or too good to rear queen-cells.

Q. Are not queens reared from the egg better than those reared from the grub?

A. I don't believe they are if the grub be young enough. Scientists tell us that during the first three days the food to the queen larva is the same as the worker larva, only in larger quantity. But it is likely the worker gets all it can eat; so theoretically a queen reared from a worker larva three days old should be as good as one reared from the egg. I think, however, that a larva of less age is better, because when bees have their choice they select one younger; I think not more than perhaps a day and a half old. Such a queen is probably as good as one reared from the egg.

Q. I wish to rear queens as soon as practicable in the spring. How can I tell when the proper time comes?

A. Not until about the time the most advanced colonies begin to start cells of their own accord. Or, to take it on another basis, not until bees are gathering enough so as to begin building comb. You can begin a good deal sooner than either of those times, but your queens will not be worth rearing, and they may have trouble in becoming mated.

Q. Which is the simplest way, rear queens in nuclei, or re-queen the selected colony by inserting a frame with queen-cells?

A. It is much simpler to hang in the hive a frame with a queen-cell, or to put in a queen-cell without the frame. Only in that case you will have to wait ten days to two weeks before the young queen begins to lay. You also run some risk that the young queen may fail.

Q. I am working my bees for extracted honey exclusively, and use a three-story hive. Can I requeen my apiary by rearing young queens in the upper story by employing two queen-excluding honey-boards, one over the brood-nest, and one under the top story in which the new queen stays? Of course, I must bore a hole in the back of the super from which the young queen can fly. Will I get rid of the nuisance of finding my young queen killed, or at least gone, when I take a notion to hunt out the old queen and decapitate her?

A. Years ago I was delighted to succeed in the way you outline, but of late years failures have been the rule, so I have given

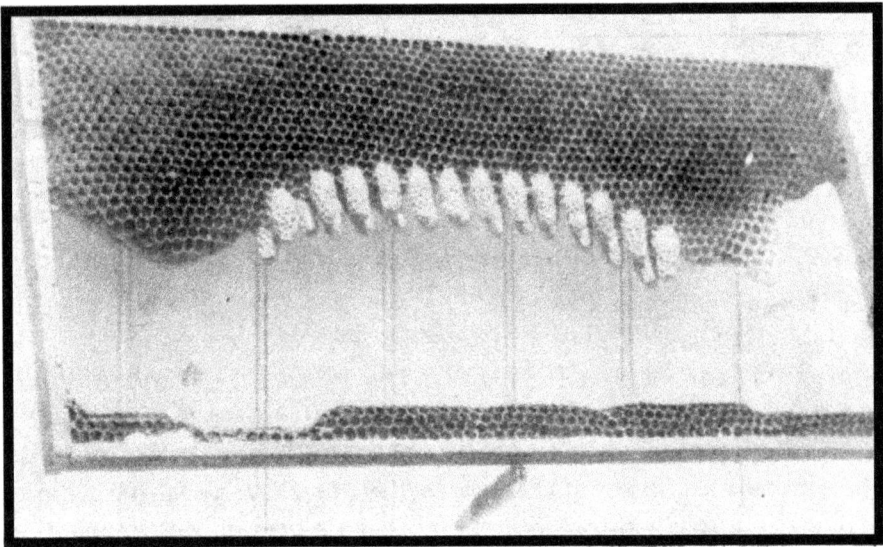

FIG. 24.—Queen-cells built on a comb specially prepared.

it up. I don't know what makes the difference, unless it be that originally the upper story with the young queen was more isolated. The farther up the top story, the better. Indeed, the first time I had a queen reared and laying in an upper story was an accident, and there was not even an excluder in the case. I put three or four stories of empty combs over a colony to have the bees take care of the combs, and in order to make the bees traverse the whole, I put some brood in the upper story—no excluder anywhere. After some time I was surprised to find a young queen laying in the upper story. The bees had reared her from the brood, and it happened that there was a leak under the cover which she could fly through. In my later attempts there has not been so great isolation, and it might be worth while for me to try again. At any rate, it is worth your while to try it, keeping in mind to have your upper story high up.

Q. I am thinking of re-queening by allowing a queen to be reared above the excluder and then allow her to come back and enter the hive below after she is fertilized. Will she kill the old queen, or be killed? If you think this is not a good plan, what would you advise?

A. If you should succeed in getting a queen reared and she should return from her wedding trip, it is uncertain which queen would be killed.-

Q. This year we had only one colony out of nine that stored any surplus honey; they were Italians in an 8-frame hive. Next year we would like to make some increase from this colony, as we have plenty of extra combs and hives.

About swarming time, if I remove the queen from this colony, in a few days there will likely be a good many queen-cells started. Now, if there happens to be cells on each frame, could I make eight nuclei from it by taking one frame of bees and then take a frame of hatching-brood and bees from some other hive, and perhaps a frame of honey, and fill up the hive with drawn combs?

Q. Yes, your scheme will work. If the cells should happen to be all on one or two combs, you can cut out a cell and fasten it on another comb by pinning over it a hive-staple. When you take the extra frame of brood and bees from some other colony, shake into your nucleus the bees from one or two more of the frames, since a good many will return to their old home. Or, to prevent returning, you may fasten the bees in the nucleus for two or three days.

· Q. How may I rear choice queens on a small scale?

A. I will give one plan that should give you the best of queens. Of course, if you rear choice queens you must have a choice queen from which to rear them. The colony containing this queen should be built up strong, if necessary, by the addition of brood and bees from other colonies, so that it shall be the first to swarm. About eight days after it swarms there should be a fine lot of queen-cells that you can utilize to the best advantage. The more nearly mature they are the better, but if left too late there is danger that some of them may be torn down by the bees. If you are willing to take the trouble, there is a plan by which you may have them fully mature. When the colony swarms, hive the swarm on a new stand, leaving the mother colony comparatively strong. You might even return some of the bees of the swarm to the old hive. Beginning about a week after the issuing of the swarm, go to the hive each evening after the bees have quieted down, put your ear to the side of the hive and listen for the piping of the young queen, which you will hear as soon as she issues from her cell. You will have no difficulty distinguishing her sharp, clear tones, even if you have never heard a queen pipe before. The other virgins in their cells will quahk in reply. Now go to the hive next morning and cut out all cells, but look sharp that none of the virgins escape which have gnawed open the capping of the cell, but are kept prisoners by the workers. In "Fifty Years Among the Bees" I have very fully detailed the way in which I rear queens for my own use, a plan I would use if I had only half a dozen colonies. I think it might pay you well to get the book just for that part alone.

Q. Is there a better way of rearing queens for an amateur without queen-rearing tools, when queens are wanted before the swarming season? If so, please explain. (Iowa.)

A. You don't need any special queen-rearing outfit for ten queens a year, nor for 100. I'll tell you how you can rear just as good queens as can be reared from your stock, with no other outfit than what every beekeeper is supposed to have on hand.

Take a frame out of the hive containing your best queen, and put in its place a frame with a starter an inch or so deep. A week or so later you will find the bees have filled the frame three-quarters full, more or less, with new comb, with larvæ well advanced down to eggs around the outside edge. Trim off the outer edge that contains only eggs, leaving the larvæ. It isn't easy to be exact about this, and it isn't very particular, only don't cut away any of the larvæ; no harm if you leave some of the eggs. Indeed, it is not absolutely necessary to cut off any of the comb; only that outer margin is in the bees' way. Now put your prepared comb in the middle of a strong colony from which you have removed the queen, and in nine or ten days cut out the cells and give them to nuclei. In about two weeks later you ought to find most of them changed into laying queens. You see, it isn't a very complicated matter, and needs no special outfit.

You note that I give no date as to when you are to do these things. I can't, because it may be three weeks later one year than another. But be sure not to begin too early. In your locality, if you were to begin in March you wouldn't get one good queen out of twenty. Figure so as to give the brood to the queenless colony when the bees are working prosperously in the fields. In your locality that probably means that the bees should not start to build queen-cells until white clover begins to yield, or any time later; and, of course, the empty frame must be given to your best colony a week or so earlier.

Q. Is the nature, quality, color, etc., of queens affected by the bees that rear them from the egg? That is, if I give a cross colony eggs from a queen whose workers are gentle, to rear a queen, will the workers of the queen reared be gentle if she is fertilized by a drone from a gentle colony?

A. It is held by some that the character of a queen is materially affected by the nature of the nurse-bees that rear her. It is certain that a young queen poorly fed will not be so good as one that has a bountiful supply of best food. That is, perhaps, the chief reason why the attempt to rear queens very early in the season is generally a failure. But take two royal larvæ, one fed by nurse-bees of the most vicious temper, the other by the gentlest of all bees, each being alike lavishly fed, and it is hard to understand that there should be any great difference in temper of the young queens, if both had the same mother.

Queens and Swarms.—Q. When a first swarm issues, how long is it before the young queen emerges in the hive?

A. The first afterswarm issues about 8 days after the prime swarm (perhaps sooner, perhaps later), and the young queen probably emerges the day before that, say about a week after the prime swarm.

Q. Is a queen on the outside or inside of a swarm which is clustered on a limb?

A. She may be anywhere in the cluster, and sometimes the bees will cluster and the queen not with them at all.

Queens, Shipping.—Q. Do you think that queens that come through the mails are as good as those not caged?

A. Certainly it would not be safe to suppose that a queen will be improved by a journey through the mails. She may not be injured at all by such a journey, and the injury may be serious. Even in a case where a queen is greatly injured by being mailed, she may be a very profitable investment. Suppose you have a strain of very poor bees, and you order a queen of a strain of bees that will store double as much as your bees. She is so badly injured in the mails that she is slow about beginning to lay, lays very sparingly, and gives up the ghost before you have had her a month. All the honey stored by her progeny, if sold at a high price will not amount to as much as you paid for the queen. If you figure merely on the honey stored by the colony in which she was introduced, the purchase of the queen was a losing operation. But that isn't the only thing to be considered. Even if she lays only a very few eggs, if you are lively about it, and from those few eggs rear enough queens to requeen all your colonies, that stock may be just as good as if the queen had never been injured in the mails at all, and as a consequence you have just doubled your future crops. In other words, the injury of a queen in the mails does not necessarily injure the stock reared from her.

Queens Stinging.—Q. Can a queen sting?

A. If you allow two queens to come together, unless one of them is very old, you will soon learn that they can sting, for one of them will soon be a dead queen. The strange part of it is that the victor is never injured in these duels. But a queen will never sting you. I have handled thousands of queens, and I never knew one of them to make the least show of stinging. Nor will a queen sting a worker. Just once in my lifetime I knew of one exception to this rule, when I saw a queen sting a worker.

Queen-Trap.—Q. As I have no time to be around at swarming time, I am going to put on the queen and drone-trap. Will that be right?

A. It will be all right if you give the proper attention afterward. But merely putting on a trap will not answer. The queen will be caught in it, and if you leave her there, there will be a young queen in the hive in a week or so, and when she tries to fly out to be fertilized she will be caught in the trap, and then you will have a queenless colony. You will have to keep watch and when the queen is caught in the trap make an artificial swarm, or dispose of matters some other way.

Queens, Two in One Hive.—Q. Wouldn't I get more honey by having two laying queens in a hive; first a hive-body, then a super, then a honey-board; next a hive-body, with the second queen; lastly a cover? Would the two laying queens fight through the honey-board?

A. The queens could not very well fight, but I don't think you would gain by the plan. One of the queens is likely to disappear before long.

Q. Will two or more laying queens in one hive prevent swarming, as told by Alexander?

A. I think the plan did not pan out very well afterward.

Queens, Virgin.—Q. How do virgin queens look?

A. A virgin queen looks very much like a laying queen, only her abdomen is smaller.

Q. How long after the prime swarm issues before the first virgin will begin laying?

A. About seventeen days.

Queen-Cells.—Q. Are queen-cells all at the end or at the bottom of the combs?

A. Bees generally build queen-cells along the lower edges of the combs. But if there is a hole, or some irregularity of surface in a comb, thus making room for a queen-cell, the bees do not despise the opportunity. In rare cases they will even build a cell separate from the comb on one of the bars of a frame. If a colony becomes suddenly queenless, they build cells over young worker-larvæ, converting them into young queens, and these cells are often built right in the center of a brood-comb where there is no hole or irregularity of surface.

Q. Which end of a queen-cell is the bottom—the end that a queen hatches out of, or the end where the egg is laid?

A. The top is the bottom, always. Sounds tangled, doesn't it? You see, it's like a teacup; when the cup stands full of tea, the bottom of the cup is toward the ground; and then when the cup is turned upside down the name "bottom" still belongs to the same part we called the "bottom" before, although the bottom now points skyward. The bees build queen-cells upside down, and so the bottom of the cell, like the bottom of the teacup when turned upside down, always points skyward. (To be sure, in rare instances, a queen-cell lies horizontally, but that occurs so seldom that it doesn't count.) Then, when we speak of the other end of the cell, the illustration of the teacup fails. For when a teacup is upside down, the part that is downward is still called the top; but the part of a queen-cell that is downward is not the top, but "the lower end." So the egg is laid in the bottom of the cell, and the young queen emerges from the lower end. Absurd way of talkin, isn't it? But please don't blame me; I wasn't born when beekeepers agreed to talk that way about a queen-cell.

Q. If the queenless bees should make a queen-cell, and place therein an egg, how long before the cell will be capped, and how long before there is a full-fledged queen?

A. In eight or nine days from the time the egg is laid the cell should be capped. But instead of an egg, queenless bees will start with a larvæ two days or so of age, and it ought to be capped four to six days later. In fifteen or sixteen days from the laying of the egg the queen will hatch.

Q. When is a queen-cell ripe?

A. The term "ripe" is applied to a queen-cell when it is near the time for the young queen to emerge, perhaps any time within two days, possibly within three, of emergence. When a cell is

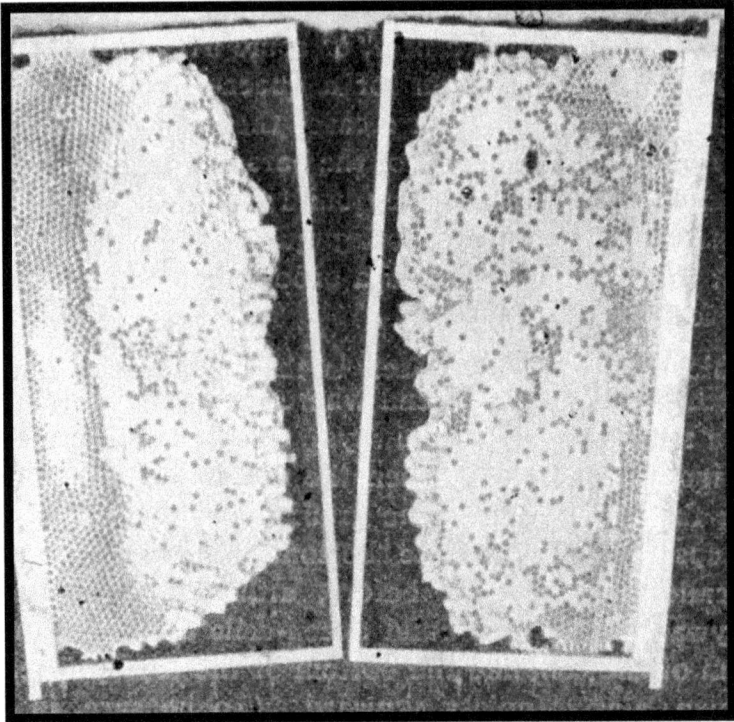

FIG. 25.—Queen-cells built on the lower side of combs by a colony which had been deprived of its queen.

sealed, you may know that at the farthest it will be only about eight days till the young queen emerges. Usually the sharp point of the cell will be gnawed away something like two days before the hatching, leaving the cell quite rounding at the end.

Q. When a colony is queenless and there are queen-cells, then one queen hatches, do the bees, or the first queen hatched, destroy the other queen-cells?

A. Both engage in the gruesome business.

Q. How often will I have to look for queen-cells?

A. There's no law against your doing so whenever you feel like it, but I suspect you mean to prevent swarming. As already said, the bees may swarm in spite of all you can do in that way, but it is hardly necessary for you to destroy cells oftener than once every ten days. If that will not prevent swarming, it will hardly prevent it to destroy them more frequently.

Q. Do you shake the bees off the combs when looking for queen-cells? If so, do you shake the bees on the ground or upon the tops of the frames in the hives? My bees seem to try to hide cells by clustering in bunches.

A. Sometimes the combs are looked over without any shaking, for if cells are in the hive at all, one is not likely to miss them all. But if a single cell is found, then it is hardly safe to omit shaking all the combs. Just how the shaking is done depends upon circumstances. If the queen is to be found, she must be found before any shaking is done, and the frame she is on set out of the hive, for if a single frame is first shaken, then it's good-bye to finding the queen. After the queen is out of the hive, the bees may be shaken on the ground, on top of the frames, or into the hive between the frames. If the queen is not to be found, the bees are shaken on the top-bars or into the hive between the frames.

Q. If I remove the queen from one of my black colonies and put in a frame with a queen-cell on it, would the queen-cell need to be protected from the bees for a few days?

A. Yes, if the cell is not protected and is given before the bees have discovered their queenlessness, it will be torn down. But in 24 hours they are likely to discover their queenlessness.

Q. Can a queen-cell, by careful handling, be cut from a comb and put into another comb for a colony, to any certainty, without injuring it in any way by pressure, or exposure, or should it always be introduced on the comb on which it is built?

A. Thousands of queen-cells have been cut from the comb and fastened in or on another comb with just as good results as if left on their own comb. Indeed, in many cases, even when the cell is taken on its own comb, it is better to cut out the cell and fasten it on the comb, for a large portion of cells are on the edges of combs where there is danger of their being chilled, and they should be centrally located, where sure to keep warm.

Q. How many queen-cells may I give one strong colony to complete during a good flow, the colony being extra strong? I mean to get good queens.

A. Some limit the number to ten. But as a colony left to itself rears twice that number very often, it is doubtful whether it is necessary to limit the number so much. Indeed, it is possible that you will do no harm to give quite a large number, say as many as 40; the bees themselves will do the limiting by destroying the excess.

Q. Why is it that queen-cells sometimes fail to hatch, even when carefully protected by prepared cages? In any case, I invariably find that the larvæ in the cells shriveled up at the hatching end of the cell; this in spite of a plentiful supply of "royal jelly," favorable climatic conditions, and during the honey-flow. I have sometimes attributed the foregoing to the fact that cells were made from a queen's first round of laying; but I have recently noted the same results with a second-season queen's brood.

A. Swiss beekeepers, who are away up in matters pertaining to queen-rearing, tell us that mere physical heat is not enough to bring young queens safely to maturity; the bees must be allowed to be in close contact with the cells, exercising some mysterious influence by their close contact with the occupants of the cells. According to that it is a mistake to cage cells as soon as they are sealed. In my own practice I leave the cells uncaged in a strong colony, not caging them till there is danger of their hatching. But there is one thing that looks as if another cause was at the bottom of the trouble. You say you find the larva "shriveled up at the hatching end of the cell." That looks as if the combs had been shaken, thus throwing the larva out of its bed of jelly.

Q. I had one colony queenless and gave it a frame of brood. The third day after, it had four queen-cells started. The next time I looked they were capped. When they had been capped about a week I looked into the hive again, intending to cut them out, but found that the bees had torn them down before it was time for them to hatch out. I also found two artificial swarms with queen-cells had done the same thing. I gave one of them another frame of brood. What is the matter with these colonies?

A. The probability is that a young queen had emerged from her cell. Her first care would be to see that all rivals were out of the way. Possibly you may say that you are sure that could not have been the case, for you looked the combs over very, very carefully, and every queen-cell was torn open at the side, the end of the cell being entire, showing that no queen could have emerged from it. In that case a virgin from elsewhere may have entered the hive. Oftener than you might suppose, a virgin or laying queen enters some other hive than her own. Sometimes, however, bees take a notion to destroy cells with no apparent reason for it.

Queenlessness.—Q. What is the best sign of a queenless colony in the spring?

A. The best sign is to find no brood present when all other colonies have started brood. Even by looking at the outside, you may be suspicious if you find the bees of a colony carrying no pollen, or very little pellets, when other colonies are carrying loads of it.

Q. Will queenless bees store honey?

A. Yes, indeed.

Q. I have two colonies that I know are queenless, and still they are busy carrying in pollen. Some say they will not do this when they have no queen. Is this so. If so, why do these carry pollen?

A. Queenless bees do carry pollen; but after they have been queenless for a time they have a surplus of pollen on hand and then they carry less pollen and smaller loads.

Q. When a colony becomes queenless, what is the best way to requeen; give it a sealed queen-cell or a frame of brood, or what would you do?

A. A cell just ready to hatch will gain about 12 days over giving a frame of brood, and a laying queen will be about 10 days better still, so if I hadn't the laying queen I would prefer the cell to the frame of brood. If it was very early in the season, I would unite with a weak colony having a good queen, rather than give a frame of brood.

Q. I have a fine 10-frame hive with plenty of stores and of bees —but queenless. I dare not order a queen from the South, as a cold snap would kill her. If a frame of brood from another hive is introduced and a queen is reared, there are no drones for her to mate with. Laying workers may develop at any time. But I do hate to lose that colony. What would you do? (March. Missouri.)

A. You are wise in thinking it best not to rear queens too early. Aside from the lack of drones, it is true that queens reared much before the time of swarming, and if drones are present, generally turn out to be so poor that they are often worse than none. All the same, you can give the colony one or two frames of brood from some other colony, with a goodly proportion of eggs and unsealed larvæ. Then within ten days kill all queen-cells started on this brood, and at the same time give a frame or two of fresh brood every ten days until conditions are right for rearing a queen, but allowing no young queen to emerge until then. You will do three things: You will keep up the courage of the colony, you will help keep up its laying strength, and you will keep it from having laying workers. And if the brood you give them be mostly eggs and very young brood, it will not cost such a great deal to the colonies from which it was taken.

After all, that's hardly answering your question, for you said: "What would you do?" In the preceding I've said what you can do. It isn't likely I'd do that. I would harden my heart and break up that queenless colony. At least I'd unite it so there would be one less colony in the apiary. If there was another colony quite weak, but with a good queen, I'd put a sheet of newspaper over the queenless colony, and set the weak colony over it. Then the bees would gnaw a hole through the paper and unite peaceably. If I hadn't a very weak colony, I'd divide combs with adhering bees among two or more colonies, taking such colonies as most needed help. In this way, although I would have one less colony, I would be likely to have more bees, and by the middle of the summer likely more colonies.

Q. If the hives are broodless and queenless by June 1, and if given a frame of eggs, larvæ, and sealed brood to rear a queen, will the queen be fairly good?

A. Young bees are the ones to rear a good queen, and in the case you mention there are probably few or no young bees, so the resulting queen would not be likely to be very good. The best thing to do with such a colony is to break it up and unite with another colony or with other colonies. If you haven't the heart to do that, then a better way than the one you mention is to give your queenless colony the queen of some other colony, and let that other colony rear its own queen.

Q. Will a colony of bees that loses its queen in October or November live through the winter? And can they be kept until May, or until the shipping season begins?

A. They are not likely to live over, but sometimes they do. It will probably be more profitable to unite them with a colony having a laying queen, even if you divide again in the spring.

Q. If you had a colony of bees quite strong with a lot of drones, that was discovered to be queenless as late as November 1, what would you do with it?

A. Before doing anything with it, I should want to be quite sure it was queenless. "A lot of drones" in a strong colony in the fall is not always sure proof of queenlessness, although something depends upon how large the "lot" is. The absence of all brood November 1 is no proof of queenlessness, neither is the failure to find the queen, for the queen is hard to find, because small. Unless you have some stronger proof than the presence of drones, better leave it till spring, and then break it up, dividing the combs and bees among your weakest colonies. If sure it is queenless, you can break it up at once. In any case, it will do no great harm to leave it till spring.

Q. If a colony of bees lost its queen in the winter, how long would it live?

A. If she were lost in the winter, the supposition would be that she laid as long as usual in the fall. The bees would become less and less in the spring, and if they did not desert the hive the last of them might be dead perhaps some time in May, or June.

Races of Bees (See also Italians, Carniolans, Caucasians, Cyprians, Punics, Etc.)—Q. Could one keep several different races of bees in the same apiary?

A. Unless it be for the sake of experimenting with a different race, it is better not to try to keep more than one kind. Even with only one, you may find it beyond you to keep them pure; for they will mix with bees as far as a mile or two away and farther.

Q. Are all breeds of bees of the same size? If not, which are the largest breeds? What is the main color of the so-called gray Caucasians? Are they gray or black, and are they as good workers as the Italians?

A. Honeybees are practically the same in size. Caucasians look so much like common black bees that you couldn't tell them apart by their looks. Opinions differ as to their gentleness and storing qualities. While some prefer Caucasians, the majority prefer Italians.

Rape.—Q. Has rape any honey value to make it worth planting for bees alone? When should it be planted to yield the most honey?

A. Rape is a fine honey-plant, but neither that or any other plant will pay to sow for honey alone, unless it be on waste land where it will take care of itself. Spring is probably as good a time as any to sow rape. In Germany rape is highly prized as a honey-plant, and sometimes bees are hauled some distance to be in reach of rape fields.

Raspberries.—Q. Do bees gather much from the blossoms of raspberries?

A Yes, the raspberry is a very important honey-plant. In some localities, notably in northern Michigan and Ontario, it yields a handsome surplus.

Records.—Q. Will you tell me some simple way in which I can keep a record of queens, etc., to see if there is any improvement in them?

A. I use a cheap blank book, giving to each colony its place, and enter there the important items, especially the amount of honey each colony yields. That matter I keep in a spot on the page by itself, so that at any time I can glance at it and tell just what the colony has done. This credit may be made in the apiary at the time the honey is taken off. For instance, this credit may appear: 24, 24, 24, 15. That means I took away 24 sections each time at three different times, and the equivalent of 15 sections at the last time, making 87 sections in all.

Redwood.—Q. I have a chance to get some hives made of California redwood. It is used for making incubators in this town. They say it will not take water, ants or moths will not work in it, and it will stand painting. Would the honey taste from it, or would the bees want to live in hives with that odor? Some say to wash the hive with salt water or peach leaves. I can get hives made from this very cheap, although redwood is high-priced. Have you had any experience in this matter?

A. California redwood has been successfully used for bee-hives, and you need not fear its effects on the honey, even without salt or peach leaves, which probably have no effect. But you will hardly find it proof against ants and moths.

Renting Bees.—Q. I let my bees on halves last summer. Should the man who took the bees have left enough stores for winter? Four of my colonies starved to death before I could attend to them, and ten more would have starved within a week if I hadn't fed them. This man told me they had plenty of stores for the winter. I let him have 28 colonies on June 5. Wasn't he supposed to return to me 28 colonies in the fall? He returned only 28 colonies with half of the increase, and the increase was eight swarms. Now, wasn't this man supposed to leave my bees in good shape, ready to put into the cellar? Is there any law on renting bees?

A If there was any law about the matter the great probability is that the law would insist on the carrying out of any contract made in the premises. So the whole thing depends upon the agreement that was made, and to make sure about it the agreement should have been in writing. If there was an agreement that you should receive back a certain number of colonies at a specified time, then the agreement should be carried out, even if the man to whom the bees were let should have to buy bees to make out the number. As to disposal of the increase, a common custom is to divide equally, but that custom is hardly law. You let the man have 28 colonies, and you say "he left me only 28 with half of the increase, and the increase was 8 swarms." If you mean by that that you got back the original number, 28, and half of the 8 swarms, or 32 in all, it would seem right. As to the condition of the bees in the fall with respect to stores, that depends upon agreement; you would get the bees back in the fall without any feeding, if the season was so poor that they needed feeding in the fall. But if they had plenty of stores for winter in the brood-chamber before turning over the bees to you, then I should say he was not trying to play fair. In a matter of this kind, if there is no written agreement, the fair thing to both parties is the fair thing to do. If you have bees and I take care of them, I furnish the time and perhaps the location, and you furnish the bees. If there is no honey, I have lost my time, and you will have to lose your bees unless you furnish me with honey to feed them, in which case I would feel compelled to do the work. But if there is a crop and I make some money, you are entitled to a part of the profit. The custom is for the man who furnishes the bees to furnish material, hives, sections, etc., and the crop and swarms are divided equally. This is a fair division between labor and capital.

Requeening.—Q. Do you think I need to requeen the colonies this fall that I requeened last summer? If they are good this spring will they be good next spring?

A. No need to requeen if the queens are good.

Q. If I introduce an Italian queen into a colony of black bees, will her offspring be pure?

A When a new fertilized queen is introduced, all the bees in the hive will be of the new stock just as soon as the offspring of the old queen have died off, and in the busy season that will be in about two months, or a little more. If the new queen is pure Italian and purely mated, then all the new workers will be Italians.

Q. To requeen all or part of an apiary with ripe cells from a breeder, I thought of dequeening about August 1 and introducing ripe cells. The bees hatched from eggs laid by the old queen from August 1 to 14 (when the new queen would begin to lay) would not, I think, aid materially in the harvest, which is about

July 10 or 15 to August 25 or September 1. How would this work? How are cells to requeen with?

A. It ought to work all right. Requeening with cells is all right, only, of course, there will be less break in the rearing if laying queens are given.

Q. I have lately bought 16 colonies of black bees. They went into the cellar on December 12, strong in bees and plenty of good, sealed stores. They are in fine condition, but as the hives are of all sizes and shapes, good for nothing but kindling wood, I shall transfer to my dovetailed hives. Do you think it would pay me to requeen them with good Italian stock early in the season?

A. Requeening early in the season is sure to interfere, at least a little, with the building up of a colony, with the possibility of interfering a great deal in case there is some hitch in introducing. If you are requeening for the sake of having better stock to breed from, it may pay to do so early, even it it interferes greatly with the honey crop. But in your case you hardly want to interfere with the crop this season. So, perhaps you will do just as well to leave the old queens until after swarming, at least, if not until near the close of harvest, unless your queens are poor, when it would be best to requeen at once.

Q. In one of my colonies I have a very prolific queen which I desire to breed from and requeen five other colonies. Later I wish to divide into two or three-frame nuclei and rear queens from this stock. Please advise the best method for me to pursue.

A. It's a bit hard to know just how to advise, there are so many ways of doing and so much depends upon circumstances, previous experience, and perhaps other things. In spite of the fact that I don't like advertising, I will say that I think that you would get information enough on that one topic alone to make the purchase of "Fifty Years Among the Bees" a profitable investment. But I'll give you one way that ought to be successful, even if you have but little experience. Strengthen the colony with your choice queen by giving it brood with adhering bees from other colonies, so it will be the first to swarm. Call it A, and name the other colonies in the order of their strength, B, C, D, E, F. When A swarms, set the swarm on the stand of A, and A on the stand of B, and put B in a new place. A week later you can cut out the queen-cells and give them to C, D, E and F, having dequeened these a day previous. If, however, you want to operate in an easier way, after you have put A in place of B, it will be strengthened by receiving all the field bees of B as they return from the field. Then it will be practically certain to swarm when the first virgin emerges, and you can leave the swarm on the same stand from which it issued, and set A in place of C. Repeat the same thing each time A swarms, setting it successively in place of D, E, and F.

Q. I am thinking of trying the following plan this season: I will find and destroy the old, inferior queen, and introduce a sealed cell (in a cell-protector) at the same time I remove the old queen. Can this be done safely? Or had I better wait about placing the cell until two or three days after removing the old queen?

A. Very likely your plan will succeed. Waiting two or three days would make the bees more willing to accept a cell, but in a West cell-protector the cell ought to be safe anyhow. The cell ought to be well advanced; then if it does not hatch out all right, it will pay to have on hand other cells so that you can destroy all "wild" cells (those that the bees start on their own brood), and give another cell of good stock.

Rheumatism.—Q. I have read that some people were cured of rheumatism by the stings of bees. I have a customer who is very fond of honey, and as she has the rheumatism badly, and is under the doctor's care, she is advised against eating honey. She was also at a Michigan bathing sanitarium and not allowed to eat honey there.

A. The fact that some people are cured of rheumatism by means of stings does not necessarily prove that eating honey is good for rheumatism. Honey and bee-poison are two very different things. Yet I have never understood that the use of honey was contra-indicated in rheumatic cases. It is possible that in the case in question some particular condition makes it advisable to deny the use of all sweets; but it is safe to say that if they are at all allowed it will be better to use honey than sugar. That able authority, Dr. Kellogg, at the head of one of the most noted sanitariums in the world, endorses the use of honey as being more readily assimilable than sugar.

Rietsche Press.—Q. Do you know anything about the Rietsche press for making foundation?

A. Thousands of Rietsche presses are in use in Europe, one reason being that so much of the foundation on the market there is adulterated. In this country there is no trouble about buying pure foundation, and although a few years ago a number had machines to make foundation, nearly all buy now.

Q. Would it pay to have a Rietsche press for 100 colonies (I use shallow extracting-frames and sections on each colony), or would it pay to sell the wax and buy foundation?

A I doubt that you could easily make foundation with a Rietsche press that would be satisfactory for section honey. For brood-frames you can make foundation with it that might be satisfactory. Whether it would be advisable to make or buy depends upon circumstances.

Robbing.—Q. What is a good sign of bees being robbed?

A. When you see unusual activity at the entrance, especially if the colony is weak, catch one of the bees that comes out with considerable bustle, kill it and see if it has honey in its sac. If it goes out with a full sac, you may count there's robbing. In a large number of the cases of robbing that occur in the spring, it is because the colonies are queenless and practically worthless, and the best thing in such case is to let the robbers carry out all the honey without disturbing them. About the worst thing is to take the hive away, for then the robbers will pitch into the adjoining hives. If you take the hive away, put in its place another hive just like it, with a comb or combs having just a little honey in them, letting the robbers clean out the little honey without disturbing the neighboring colonies.

Q. If bees begin robbing a hive, can it be stopped, and how?

A. If bees have a good start at robbing a weak colony, it is a hard matter to stop them. Perhaps the best thing is to take away the colony, putting it down cellar for two or three days, and put in place of the hive another hive like it, containing some comb and a little honey. (If you leave nothing for them to work at, they will attack one or more of the nearest colonies.) When they have cleaned out the little honey, and satisfied themselves there is no more to be had, they will quietly give it up. Then, after two or three days, return the colony to its place, closing the entrance to a very small space, perhaps allowing passage for only one or two bees at a time, and it may be that the robbers will not make another start, especially if a good queen is present. But if the colony is queenless, the case is hopeless. Sometimes robbing has commenced at a fairly strong colony with a good queen. The first thing is to limit the entrance. Perhaps painting carbolic acid about the entrance will answer. A pretty good way is to pile hay or grass in front of the entrance and keep it well wet with water.

Generally robbing is owing to some carelessness on the part of the beekeeper, and prevention is better than cure.

Q. (a) Last summer I cut a bee-tree and secured a fine swarm of Italians, with a fine-looking queen. I put it in an 8-frame hive, and in a few weeks examined it, and it had six frames of capped brood, and the other two frames very nearly full. In a few days I noticed the bees dragging out their young, and every morning the ground would be covered with young bees not large enough to fly. I opened the hive and found they were tearing the combs to pieces and had nearly all the brood out of the combs. The queen was still in the hive and seemed to be in good condition. In a few weeks more I opened the hive again, and found only a handful of bees—queen and bees had disappeared. They were within a few feet of the kitchen door, and I do not think they could have left without some of us hearing them. Can you tell me what was the matter?

(b) Do you think it will be safe for me to use these frames of comb in another colony this spring?

A. (a) I don't know. The only way I can account for the combs being torn is that robbers did it. They might also drag out the young bees, leaving the queen, at least for a time. Yet it seems very strange that a colony strong enough to have six brood-combs should have been overcome by robbers. Perhaps they were starving.

(b) If my guess is right that the combs were torn up by robbers, then it will be safe to use them again.

Q. I have read in the bee journals about bees that seemed determined to rob, and if any of them are that way probably I have some of that stock. I would be glad if you can tell me where I can get a stock that is not inclined to rob.

A. It is possible that there may be a strain of bees naturally given to robbing; yet you will find that all bees are inclined that way when opportunity offers at a time when nothing is to be had in the field. Please understand that bees have no morals, and when they can't get honey from the fields it seems entirely honest to get it from some other hive if they can, and you will probably find that the better they are at gathering from the field the better they are at robbing if they turn in that direction. When robbing occurs, it is not generally because the bees are such bad robbers, but because the beekeeper has done some fool thing to expose a weak colony and start robbing. Keep colonies always strong and avoid the start. Bees that have once engaged in robbing are the more inclined to begin another time, but it is not true to say of them, "Once a robber always a robber."

Q. This last summer, after the honey-flow was over, I noticed a lot of robber bees prowling around, and every now and then one would manage to slip past the guards and steal a load of honey. Finally they overpowered one and came very near robbing it before I got them stopped, and I got them perfectly quiet at one time and contracted the entrances to all the colonies. In a few days there came a little rain, and after it cleared up they started to prowling around again, and kept it up until cold weather, but they were worse after a rain or damp spell than at any other time. Is that their natural way of doing, or should they keep quiet during a dearth of honey, and would it have resulted in a general case of robbing if I hadn't contracted the entrance?

A. It is a common thing for bees to prowl about and try all crevices of hives at any time when gathering has stopped, and after a rain, and it is quite possible that your narrowing the entrances may have prevented a bad case of robbing.

Rocky Mountain Bee Plant.—Q. Please give a description of cleome. If planted in the spring, will it flower during the summer? What soil is best for it? Is it an annual? What time in the year should it be planted?

A. Cleome integrifolia, or Rocky Mountain beeplant, grows wild in some parts of the West in large quantities, and is an excellent honey-plant. Some years ago it had quite a boom, and seed was planted largely. But it is doubtful that anyone who sowed seed ever got back the cost of the seed. A. I. Root found it inferior to its near relative, the spider plant, although neither was worth cultivating. It is doubtful that it is worth while for you to try it if it does not grow wild with you. It should be sown in spring; is an annual, so blooms the first year. I don't know what soil suits it best. I know it does well here in good garden soil; but the acres of it I saw out West were growing wild on land that looked to me poor.

Roofing Paper.—Q. Please discuss the advantages and disadvantages of roofing paper, such as "Ruberoid," as a wrapping paper for the winter protection of bees.

A. I don't know enough about "Ruberoid" to discuss it fully, yet if you mean to use it to wrap about the hives in winter, I should think it altogether too heavy, if it is the same material that is used for roofing. It has the advantage of durability, and for covering over the top is no doubt excellent. But its heaviness, and especially its stiffness, would make it unfit to wrap about a hive to be tied on. The lighter tarred building paper is better for that, and less expensive. But I am not speaking from experience, as I winter bees in the cellar, where nothing of the kind is needed.

Royal Jelly.—Q. How long will royal jelly, taken from a queen-cell, keep and still be fit to use in grafting cells?

A. Much depends upon the thickness of the jelly and how open it is kept. If very thick, in a warm place, with air stirring so as to encourage evaporation, it might be unfit to use in less than an hour. Not very thick, in a cool place, with little chance for evaporation, I guess it might keep two or three days.

Sage.—Q. Where does the sage honey come from?

A. California. It is one of the principal honey-plants of that state. Sage honey is of very fine flavor and finds ready sale in any market. There are several varieties of sage growing wild in California, all of them yielding more or less honey.

Salt.—Q. How is salt fed to bees?

A. It is not often that salt is fed to bees. Some have thought it desirable, because in the spring bees are found where salty moisture is to be obtained. Others think the bees care only for the moisture, and prefer a place not because the water is salty, but bcause it is warmer than in other places. The easiest way to give salt to bees is to give it in their drinking water

Sap.—Q. Is sap from rock maple good feed for bees? If so, should it be boiled down or fed as it comes from the tree? If boiled down, how far should eight quarts be reduced to make the best feed?

A. Yes, the bees will take it without boiling down. But look out not to feed it on days too cool for bees to fly freely.

Scent in Bees.—Q. Do bees have a sense of scent? If so, where is it located?

A. Yes, bees have a sense of smell. Until lately this sense has been believed to be located in the antennae. Now Dr. James A. Nelson locates it in different parts of the body.

Sealed Covers.—Q. I am a young beekeeper. I had seven colonies last fall, and put all of them under cover on the south side of a shop. I thought they would be good and strong in the spring, but when I set them out I found only two alive. These were good and strong. I had sealed covers on them, but they looked as if they had been pretty damp. They all had plenty of stores except one. Do you think the dampness killed them? What plan would you suggest for me to take next winter?

A. Likely the dampness had much to do with it. With only a single thickness of board for a covering, it would get quite cold, and the moisture from the bees would condense on it and fall in drops on the bees. To avoid this, have a super or some kind of a rim over the hive, and have this filled with rags, crumpled newspapers, planer shavings, or something of the kind; this filling resting on burlap which is directly over the frames. Even with the covers just as you have them, you could pile a lot of packing on top of the covers, and this would help a great deal, for it would make the sides of the hive colder than the cover, and the moisture would settle on the sides instead of the top. It would be a good plan for you to find within ten or twenty miles experienced beekeepers who winter successfully, and find how they winter.

Sections, Clean.—Q. How would you keep the section-boxes white and clean? The sections I took out were all covered with propolis, and were a sorry looking sight. How can this be helped?

A. In some supers the sections are protected so that the bees cannot get at much of the wood to soil it, but with the best that can be done they will be able to get at some of the wood, and the bees are sure to crowd glue into the cracks that must be made by covering up, for it is their nature to crowd glue into any crack not big enough for them to crawl through, while a planed surface fully exposed will get very little glue. I prefer T-supers which leave bottom and top of the sections entirely exposed, and then they are scraped with a steel cabinet scraper and sandpapered.

Sections, Granulated Honey in.—Q. I had a large number of partly filled sections last season, and the honey granulated before I found time to extract it. Can I put these sections into the supers in that condition, or would you advise setting them out for the bees to clean out before using?

A. Don't think of putting them on again unless you can have the honey cleaned out of them thoroughly by the bees, and next time have that done in the fall.

Sections, Kind to Use.—Q. As the price of sections is very high, and section lumber very cheap in my neighborhood, do you think it would pay me to buy a machine to make sections?

A. No; there is probably not a man in the country who makes sections only for his own use. A complete outfit of section machinery would cost several thousand dollars, probably.

Q. Do you prefer beeway sections to plain, and why?

A. I prefer beeways because they are more easily handled

without danger of thrusting the fingers in them. Although I might never jam my fingers into a plain section, the danger comes when the retail grocer handles them. But even while in the bee-keeper's hands, they must be handled with more care, which means more time. A plain section tumbles over more easily than a bee-way section. A plainer, cheaper separator goes with the beeway. On the other hand, it takes a smaller case for plain than for bee-way sections, although it's easier to lift the beeways out of the case.

Q. In using beeway sections, do you put the beeway at the bottom, or at the side?

A. The beeways are to allow a way for the bees to go up, so they are at the top and bottom of the section.

Q. Is there any advantage in the sections open four sides, or open top and bottom only, or only on bottom? If there is any, what is it?

A. At one time it was claimed that with the four sides open the bees would have more free communication, and would fill out the capping to the wood better. The few that I tried did not seem to have this advantage. A section open top and bottom is absolutely necessary if you tier up. If you never use more than one super at a time the opening at the bottom would be enough.

Q. (a) What style of section would you advise? I had thought of the ideal, 3⅝x5x1½, plain.

(b) Are not many of the best plain sections ruined for ship-ping by the bees drawing them a little beyond the wood?

A. (a) After trying more or less the different kinds of sec-tions, I settled down some time ago upon the 2-beeway sections, 4¼x4¼x1⅞. I think this is the preference of the great majority of comb-honey producers.

(b) I don't think there is so much objection on that score as there is because the plain require so much more care in handling lest the fingers be thrust into the comb when they are handled. More care must always be taken in setting down a plain section, lest it topple over. During cleaning, the plain section is more likely to be injured. In general it may be said that the projection of the wood in a beeway section is a protection, although it has more of a lean look than the plain.

Q. Which section is better for the 8-frame super, the 4¼x4¼x 1⅝ or the 4x5x1⅜ section?

A. There is little to choose, but most beekeepers would prefer 4¼x4¼x1⅞ to either.

Q. What about two-pound sections?

A. They had the field when sections were introduced years ago, but side by side brought 2 cents a pound less than the 1-pounds; so that notwithstanding the less labor in their production, they were driven out of the market. It is somewhat doubtful whether they would do any better now.

Section-Folding.—Q. How should I manage the sections? Must they be wet before bending, or bent dry? I see a hand-machine advertised for bending them. Would you advise the use of one, or bend by hand?

A. Sometimes sections can be put together all right without wetting; generally too many of them will break unless the joints are wet. If you have many sections to fold, you will find it better to have some kind of section-press.

Sections, Short Weight.—Q. It rather displeases me that my sections of honey (1912), while looking well, averaged in weight only 13 ounces, while about all the others I weighed in this neighborhood weighed at least 14 ounces, and sometimes more. Are there any reasons evident for such discrepancy? In 1912 I had five times more honey than in 1911.

A. You will probably find that in flush years, when honey comes in rapidly, combs will be filled out more plumply than in a slow flow, perhaps because in a slow flow the bees have more time to build wax and seal combs. You will also find that they will fill combs more plumply if crowded for room. Like enough you gave the bees more surplus room than your neighbors did. Taking one year with another, you are probably the gainer by it.

Sections, Taking Off.—Q. Do you take off sections as fast as finished, or do you leave all on until the flow is over?

A. Neither. I take off each super as soon as it is finished, or finished all but a little at the outside corners.

Q. What is the best way to get the bees from the sections when I remove the super? If one should take the super a distance from the hive and brush them off, would they go back to the hive, or would they get lost?

A. The Porter bee-escape is a nice thing to use if you have time to wait for it; and if you want more prompt work, there is, perhaps, nothing better than the Miller tent-escape, which latter you can make yourself. It is probable that you do not have enough honey to make it worth while to have an escape; yet I think if I had as many as five colonies I should make a Miller escape. Without having an escape there are several ways to proceed. The way you speak of will work, for if you brush the bees off close to the hive or several rods away, they will find their way home again, unless there be some bees on the section so young that they have never left the hive—a thing not likely to happen. Another way is to pile up several supers in a pile, bee-tight at the bottom, and over the top spread a sheet or other covering that is bee-tight, but will let the light through. From time to time lift off the sheet and let the bees that are above escape, and in the course of a few hours all ought to be out. Whatever way you do, it is well to smoke down a good part of the bees before removing the super; but don't be too lavish with your smoke or the honey will taste of it, and smoke doesn't improve honey as much as it does ham.

Sections Unfinished.—Q. What causes bees to leave a few sections of honey uncapped in the central part of a super all filled with honey; plenty of bees and warm weather?

A. There's a difference in nectar, some of it being ripened up more slowly than the majority. It is just possible that when the bees commenced work in the supers, the central sections were filled with nectar of this kind, or with honey that possibly for some other reason they were slow about sealing, and then the rest of the super was filled with honey of a character to be promptly sealed. Another possibility is that the central sections were in some way objectionable, possibly from having foundation or comb that had been used before and left too long in the care of the bees when not being filled, and so covered to some extent with propolis. Still another possibility is that there was brood in the central sections; then, after the brood hatched out they were filled with honey which, of course, would be later in being sealed. Another possibility is that this was drone-comb and the bees left it without honey for a long time in the hope the queen would find it and lay eggs in it.

Q. If I use your plan of taking off honey, taking the filled and capped sections, are the unfinished ones returned to the same hive and in place of the ones taken out new sections put in, or do you fill this super with other partly-filled sections taken from another hive?

A. The unfinished sections from different hives are assembled into one super, and then this super is put back, possibly on a hive from which none of the unfinished ones were taken, no attention being given to where the sections came from.

Q. What do you think of putting down into the brood-chamber say, perhaps two framefuls on each side?

A. The plan is not used so much as formerly, if indeed it is now used at all. If you leave the sections below to be finished, you are likely to have pollen in them, and also to have the cappings darkened. When sections have been put below it has generally been merely to get the bees started in them and then put them up. But it is not advisable to put them below at all.

Q. Will partly-filled sections do for fall feeding in place of sugar? Would they keep until next year if properly cared for?

A. They will do nicely for fall feeding, and will keep well for use the next year, or for five years later, after the bees have cleaned them out in the fall. But unless the honey is thus cleaned out in the fall, it is not likely you can keep them so as to be used the next year.

Q. I have a lot of sections that were on the hives last season, but owing to the drouth, which caused a sudden stop in the honey-flow, they were not completed. Some of them contained some honey, which I allowed the bees to remove last fall, and merely started to be drawn. (They had had full sheets of foundation in the first place.) Shall I use these as they are this season or will there be too much mid-rib to make the best honey? I have often used the "go backs" for baits, using from one to four in a super, but I have 20 to 30 supers full now.

A. If you have unfinished sections that are fall-emptied and in good condition, use them and be exceedingly thankful for every one you have. Bees do not add to the mid-rib, no matter how long sections are left on the hive; but there is danger if they are left on too long in the fall that the bees will plaster them over with propolis, in which case there is nothing to do but to cut out and melt them up.

Q. I would like to learn of a good plan to clean up sections that are unfinished in the fall of the year. I have thought of tiering up supers 12 or 15 high, and let the bees rob them out, but as my yard is close to the house and buildings, I do not like to do it, and I don't know which is the better way. I have at present 105 colonies, and expect to increase to 175 this season. I had about 2,000 of these sections, and am using them tor baits, and find them excellent in starting the bees to work in the supers.

A. I have had much experience in getting the bees to clean up sections in the fall, and have found no better way than to let the bees rob them out. If you pile them up, as you suggest, allow an entrance large enough for a single bee for each 5 or 6 supers.

If the bees can get at them more freely, they tear the combs to pieces. With a sufficiently large number to be cleaned out, say something like a super for each colony, you may go to the other extreme and spread them all out so as to let the bees have free access- to the whole business at once. I spread the supers about in my shop cellar, and when all are ready I open the door and invite all bees to help themselves. They are protected against rain, and may remain several days until the bees have them thoroughly cleaned out. If you pile them up on top of hives they will be cleaned up, but the bees are likely to put some of the honey back into the sections. Someone, I think, has reported success by piling supers back of a hive, allowing access by way of the bottom-board without allowing other bees access. I never tried it. I have tried putting them in front and it was a failure. It might work better behind.

Selling Bees.—Q. When is the best time to sell bees, to get the highest prices?

A. In the spring.

Separators.—Q. Do you approve of the separators in supers? I have two supers of honey before me now, just taken off, one with and one without the division-boards, and I find the bees have fastened the sections nearly all over to the board.

A. I most certainly approve of separators for sections that are to be packed for shipping. If the sections are intended for home use, it is as well to have no separators. I have no trouble with sections being built to separators, but hives are level from side to side. Bottom starters help, too.

Q. What separator do you consider best, fence or sawed wood? I ordered sawed wood, slotted top and bottom.

A. All things considered, I prefer to use a plain wood separator, sliced or sawed, with no slots or scallops.

Sex of Bees.—Q. Does the size and shape of the cell in which the bee is reared have anything to do with the kind and sex of the bee? Or is it the food on which the larva is fed that determines the sex and kind of bee, as the eggs that bring forth the three kinds of bees are all laid by the one queen?

A. The sex of the bees depends upon whether the egg is fertilized or not. An unfertilized egg produces a drone, a fertilized egg a queen or a worker. An unfertilized egg in a worker-cell can produce only a drone; a fertilized egg in a drone-cell can produce only a worker or a queen. Under normal conditions only unfertilized eggs are found in drone-cells, and fertilized eggs in worker and queen-cells. The egg that produces the queen is practically the same as the one which produces a worker; only the cell is enlarged, and the bees feed it throughout its larval existence the richer food that is given to the worker larva during its first three days.

Shade for Bees.—Q. Can bees be given too much shade in early spring?

A. Yes; at that time it is better to have the sun shine on the hive at least part of the day.

Q. Is it very necessary that a colony have shade during the heat of the day?

A. Different views are held as to the desirability of shade for bees, some even saying that they are better without it. No doubt there is in this respect a difference in localities. In my own locality I think they are better off with some shade.

Q. Is there any danger of losing swarms if the hives are in too hot a place? Is it necessary to keep bees in the shade all the time?

A. There is great danger that a newly-hived swarm will desert if the hive is too hot. After it becomes settled and has started brood, the danger disappears, and a colony may do well without any shade. Yet in most places it is better that a hive shall be shaded in the heat of the day. A nice thing is to have a hive under a tree which shades it in the middle of the day, but allows the sun to shine upon it in the morning and evening.

Shake-Swarming.—Q. What is meant by a shaken swarm?

A. When the bees are shaken or brushed from their combs, and all the combs, or all but one of them, are taken away, that is called shaking a swarm, and the bees left in the hive are called a shaken swarm.

Q. Last year I had a lot of trouble with runaway swarms. Can you tell me how to practice "shake swarming"?

A. Lift the combs out of the hive, one after another, and shake the bees back into the hive, filling up the hive with empty combs or empty frames and when you have done that you have shaken a swarm. Of course, you must be sure that the queen is left in the hive from which the brood has been taken. You can make any disposition you like of the frames of brood taken away. They may be used to strengthen weak colonies, or you can use them to make new colonies. If used in the latter way enough bees must be left with them so the brood will not be chilled, unless you live where it is so hot that there is no danger of chilling. The more bees, however, you can leave with the swarm, the better work it will do on surplus.

Q. I would like to practice the shake-swarm method. What would be best to shake the bees on, empty combs, starters, or full sheets of foundation?

A. Empty combs are probably best, and full sheets of foundation next.

Q. It is impractical for me to stay at home and watch for swarms, so I must resort to artificial swarming or dividing— probably the brush-swarm plan. At what stage of queen-cells should the swarm be shaken, when queen-cells are started without brood in them yet, after brood can be seen in them, or after they are capped over?

A. Swarms may be shaken without paying any attention to queen-cells as soon as the season of swarming comes, or as soon as colonies are sufficiently strong. Some prefer to wait until a number of cells are found containing eggs or larvæ. It would hardly do to wait till sealed cells are present, for at that time a swarm is likely to ensue. The presence of queen-cell cups with neither eggs nor larva in them need not be considered, for these may be found at any time, even in winter.

Q. How far should a shaken swarm be set from the parent hive?

A. A shaken swarm is left on the old stand.

Q. In removing the old hive to a new location, and putting a new hive on the old stand, is it essential that the old queen should remain, or be shaken into the new hive on the old stand, or can she be put into the old hive on the new location?

A. The queen is to remain on the old stand with the shaken swarm. The point is that the brood is to be taken away.

Q. How did driving on capped-brood work with you in the control of swarming and securing honey?

A. So far as I could see, giving sealed brood to a driven swarm worked just as well as giving foundation, and of course made a stronger colony.

Q. I practiced the shaken-swarm method a little last summer, but some of them would swarm out again the next day. What was the cause of that?

A. Possibly it was hot in the empty hive, and they swarmed out just as a natural swarm often does under the same circumstances. A frame of brood may hold them.

Shipping Bees.—Q. Will shipping bees cause queens to be drone-layers?

A. When queens are shipped by mail in cages, it is not a very uncommon thing that they are somewhat affected as to their laying, but I don't remember that I ever heard of a queen being made a drone-layer by it, and I don't think I ever heard of a queen being seriously affected when shipped in a full colony.

Q. Do I have to have my bees inspected before I can ship them? They are free from foulbrood.

A. It depends upon the laws of the state into which you ship. Some states require inspection, and others do not. Whether they have foulbrood or not has nothing to do with the case; if the law requires inspection, all bees shipped into the state must be inspected. Better have them inspected as a safeguard.

Q. When is the best time to ship bees in the spring?

A. In freezing weather the combs are somewhat brittle, and likely to break easily, and the bees do not stand a journey as well as when more active. When combs are filled with honey they are likely to break in transit, and if too warm there is more danger that the bees may suffocate. So the best time in spring is while the combs are mostly empty, any time after it is warm enough for the bees to fly nearly all day.

Q. I am considering shipping bees in the fall, say 100 miles or more, and then giving them a good flight before putting them in the cellar. Do you think it injurious to their wintering well to ship them in the fall?

A. If they have a good flight before being taken in the cellar, I should not expect any harm from the journey. The excitement of the journey, however, would make them eat a little more, so you would have to be a little more careful to see that they had stores enough.

Q. How would you advise packing a colony of bees which is to be shipped by freight to New York from Texas in early spring?

A. There must be good ventilation, and everything as firm as possible. You can make sure of the first by having a cover entirely of wire-cloth. Under this it may be well to have a sponge filled with water. If the frames are loose-hanging they must be made secure, either by nails driven down through the ends of the top-bars or by spacing with sticks. Put on written instructions for the hive to ride so the frames shall be parallel with the rails, a hand or an arrow to point toward the engine.

Q. I want to take a few swarms with me to Minnesota about July 30. The car will likely be on the road about a week. How shall I prepare the bees for shipment? (Illinois.)

A. The frames in your hives must be fastened so they cannot move about, although that is not necessary if you have frames with fixed distance, as you probably have. If the entrances to your hives are two inches deep, closing them with wire-cloth may give all the needed ventilation. Otherwise better have the entire top covered with wire-cloth by means of a frame an inch or two deep. With only a few hives, you can have each one on the floor, kept in place by cleats nailed onto the floor. If the weather is very hot, sprinkle the bees with water every little while.

Q. How about fixing bees in the hives to be shipped 1,500 miles by rail? How shall I go about it to do a good job, so there will be no bees getting out, and how should they sit in the car—lengthwise?

A. To make a good job of it is something of an undertaking, In brief, you will use wire-cloth for ventilation, having the entrance closed with it, and having a frame the size of the top of the hive covered with wire-cloth, which frame you will fasten upon the top of the hive with four wood-screws. You will put the hives in the car with the frames running in the same direction as the rails of the railroad, nailing strips on the bottom of the car so the lower tier of hives cannot shake about. You must not set the upper tiers of hives piled up directly upon one another, for that would stop ventilation; but over the lower tiers you will put 2x2 or 2x4 scantling, running across the car, on which to rest the upper tiers, thus leaving a space for ventilation. You will probably use a cattle-car, which favors ventilation; and you will see to it that you can get at all the hives to spray the bees with water when they become excited and heated.

Shipping-Cases.—Q. Do you get shipping-cases returned?
A. No.

Shipping Comb Honey.—Q. I write for a little information in regard to shipping honey to Chicago, or other large cities. Is it necessary to enclose the shipping cases in extra strong boxes, or will they stand the rough handling without extra casing?

A. If section honey is sent in shipping-cases without any outside protection there is danger that it may not go safely. No need to put the cases in heavy boxes that are close. Crates, or carriers, as they are called, should be used, which are more or less open, only close enough so they will hold the cases, the object being not so much to cover the cases as to prevent rough handling. If cases are shipped without being in carriers, railroad hands are likely to throw them as so many bricks, putting them in the car in any sort of position. Years ago I shipped some cases loose in a car, to go a pretty long distance, and when they were transferred to another car some of the cases were on their sides, and, of course, badly smashed sections of honey was the result. A carrier is generally made to contain eight 24-section cases, or sixteen 12-section cases, and provided with handles. Being so heavy, they are necessarily handled with less roughness than would be the loose cases. Load in car so sections run parallel with the rails. If sections are sent in cases, unprotected, they take a higher freight rate than if packed properly with a layer of straw on the bottom.

Shook Swarming.—Q. What do you call "shook swarming?"

A. "Shook swarming" is bad English that has, I am sorry to say, grown into quite common use in place of "shaken swarms," or "shake-swarming." Perhaps a more appropriate name would be one used in Germany, "anticipatory swarming." (See Shake-swarming.)

Smartweed.—Q. Will honey gathered from smartweed be strong in taste like pepper? Last year the honey gathered in the fall was so strong after being swallowed that it would burn the throat for two or three hours. Smartweed was plentiful.

A. The general run of what is called smartweed honey will not smart your mouth at all. But the plant from which it is gathered hardly ought to be called smartweed, for if you chew the leaves it will not smart your mouth any more than to chew so much lettuce. It also goes by the name of heartsease—the better name, the botanical name being Persicaria. Persicaria punctatum is the real smartweed, and if you chew a leaf of that you'll wish you had let it alone. I don't know about the honey from this, whether it is acrid or not, but it is possible.

Smoke.—Q. When is the best time to blow smoke in at the entrance when opening a hive, on a cloudy day, or sunshiny day, or both?

A. The time to blow smoke into the entrance is just before you take off the cover, no matter what kind of a day.

Q. How long can you keep the hive open when handling bees without smoke? When they come to the top of the frames do you smoke them back?

A. Maybe one smoking will do for all day; maybe two minutes. So long as the bees remain peaceable they need no more smoke. No matter if they do come to the top of the frames, so long as they remain good-natured, but when they begin to fly at you, give them enough smoke to make them behave.

Q. How much smoke should I give a cross colony of bees?

A. As in other cases, use as little as possible, but enough to subdue them. That isn't very definite, is it? Perhaps I might say keep on smoking so long as the bees keep darting out at you, and stop as soon as they beat a retreat. With most bees a very little smoke is necessary, and if you keep on smoking they will boil out and run over the sides of the hive at the top. That's too much. Some have reported bees of such disposition that smoke seemed to have little or no effect on them, if, indeed, it did not make them fiercer, and the only thing to do was to manipulate very carefully. Yours are hardly of that kind.

Smokers.—Q. What size smoker is the best to use?

A. I never saw a smoker too large, although with only one or two colonies you can get along with a small one. Sometimes you want a bigger volume of smoke than a small smoker will give, and you can use just as little smoke as you like with a large smoker. The large smoker holds fire better than the small one, and you can more easily have fuel to fit the large one.

Q. At what time do you consider it necessary to use the smoker? Do you think too much strong smoke is injurious to brood and queen? I prefer using brush and veil as much as possible. I am stocking up with selected Italians. Are not these the most gentle bees we have?

A. If I were keeping bees merely for the fun of it, I might handle them without any smoker at all, and with very gentle ones it may never be absolutely necessary. Aside from quieting the bees, no good can certainly come from blowing smoke into a hive, and no more should be used than necessary. But as a matter of actual practice I generally give a puff at the entrance before opening each hive, and a little over the top as the cover is removed. I can hardly afford the time to go slowly enough without any smoke. Bees are like folks—they differ in disposition. Italian bees are very gentle in general, but there are exceptions. As a whole, they probably do not excel the Carniolans in gentleness.

Smoker Fuel.—Q. What is the proper fuel for the smoker?

A. It s largely a matter of convenience. Any old thing that will burn is likely to answer all right, provided it is easily obtainable. Probably nothing is better than dry hardwood chips. A favorite with some is a greasy cotton waste that is thrown away after being used in machine shops or on automobiles or locomotives. Then there is bark, planer chips, cowdung, etc.

Q. So far as using a bee-smoker is concerned, I am a novice in the business, and with poor success so far. I have tried newspaper, excelsior, and cobs broken up, but by the time I have the cover off and begin to raise the inside cover over the hive, or sections, the fire is out. What is best to use to make smoke? How should I use the smoker to keep the fire from going out?

A. Try old rags of any kind of cotton cloth. First put a little loosely in the smoker; light it and let it blaze up; put a little more on and keep blowing till it gets a good start; then fill up and it will not be likely to go out till it burns out. When the smoker is not in use it will burn better to stand upright. Almost anything is better than newspapers.

Q. Explain how you prepare saltpeter rags for smoker-fuel.

A. I take a 2-gallon stone crock, perhaps half full of water, put in it half a pound or less of saltpeter, fill up with cotton rags, lift the rags out and let the water drain back into the crock through a colander or leaky pan; spread the rags out on the grass for the sun to dry, and they're ready to use. I use them only to start the fire, filling up the smoker with hardwood chips or some other fuel.

Snow.—Q. Can I put snow over the entrances of the beehives when the coldest, windy days come?

A. It will be all right if the snow is dry. If the snow is wet and packs together it may smother the bees.

Q. Is it necessary to keep snow and ice swept away from the entrance of hive?

A. As long as it remains dry and hard, a little snow at the entrance is not likely to do any harm. But if it becomes wet and soft, filling the entrance and then freezing, it may do harm, so it should be cleared away before it has a chance to freeze. Not that there is special harm from the freezing, only that it allows the entrance to remain closed.

Q. My colonies are buried under the snow. Will they smother? I have planer shavings on top to let the air through.

A. Enough air works in through the snow so there's no danger of smothering. But look out not to let the entrance become filled with damp snow and then freeze solid.

Spider-plant.—Q. I notice in reading that there is a plant called spider-plant. Will it grow here? What time in the year should it be planted? (N. Carolina.)

A. The spider-plant will probably grow with you if sown in the spring, but you will hardly find it worth the trouble.

Spiders.—Q. Do spiders ever injure bees?

A. Not to any great extent. If their webs are allowed at the entrance of a hive, a few bees will be caught and killed.

Spraying.—Q. A man living a mile from me is going to spray his apple trees with Paris green this spring. Will my bees bring it home to their hives?

A. If fruit trees are abundant, your bees may not go so far. If scarce, they will be likely to visit those trees when in bloom, and if he should spray during bloom it would mean death to the bees. But if he is an up-to-date fruit-grower he will spray only before and after bloom. The experiment stations have clearly settled that spraying during bloom is a damage to the fruit crop, and in several states it is against the law to spray during bloom.

Q. Have you ever had any serious loss from poison being used to spray apple trees? (Illinois.)

A. I think not; but I have some trouble with spraying in time of cherry bloom. The owner of a large cherry orchard is one of the best and straightest men in the community; but somehow he can't get it through his head that he is hurting himself by spraying when trees are in bloom, and he says if he doesn't begin spraying a little before the bloom falls, that he can't get through the whole in time. When as good a man as he is cannot see any wrong in subjecting me to serious loss for the sake of a little inconvenience to himself, it shows that no effort should be spared to have Illinois come to the front like some other states, with a strict spraying law. If I understand the matter rightly, a man lays himself liable to penalty if he puts out poison purposely to kill my bees, but if he poisons them accidentally while spraying fruit bloom he goes scot-free.

Q. I am a beekeeper in a small way, having 64 colonies; but I am going to have a hard struggle, as people here spray when the bloom is on as well as when there is none. Spraying fruit trees is the thing, but not when the bloom is on. It doesn't do any good to talk to people. If we haven't any law in the state, why don't the beekeepers go together and get a bill before our legislature against this spraying when the bloom is on? That is all that will ever stop it.

A. You are quite right about the importance of a law against spraying, but I am sorry to say there is no law upon the subject in Illinois. A few years ago quite an effort was made in that direction, and a bill introduced in the legislature, but it was buried in committee. The trouble is that the chairman of the committee to which such bills are referred has always been a fruit-man rather than a bee-man. I wrote to the chairman of the committee at that time and he replied that fruit-men all knew that spraying fruit trees when in bloom was against their own interests, and so there was no law needed. Of course that looks reasonable; it would seem hardly necessary to have a law against a man building a bonfire under a live apple tree. All the same, there are ignorant orchardists, as in your neighborhood. I think that the real milk in the cocoanut is that those fruit-men, who are none too conscientious, although they know it to be against their interests to spray during bloom, want to spray as near that time as possible, and don't want the risk of butting up against the law if they should happen to spray at a wrong time.

Spring Protection.—Q. (a) I wish to give my hives spring protection. I have read of roofing-paper or felt being used, but would not this plan do? That is, just get a dry goods box (which can be done very cheaply), and both sides and bottoms being made of matched lumber, take the top off, and invert the dry goods box over the hive, a sufficient hole for entrance being cut into the dry goods box to open over the hive-entrance? I figured on no packing of any kind between the box and hive.

(b) Would not ordinary oat straw covering over the hives, leaving an entrance, be good spring protection? Or would a wet spring keep the hives too damp?

A. (a) Your plan may work, but the trouble is that when the sun shines out for a short time, or even for a long time, the bees don't get the benefit of it, the dry goods box keeping them cool; whereas, with the black felt they would be made warmer than with the naked hive.

(b) It will be all right if some covering to shed the rain is over the straw, or if the straw is so placed as to shed the rain.

Splints.—Q. What are your splints, for staying foundation? How are they used?

A. They are splints one-sixteenth of an inch square, of basswood or some other straight-grained wood, about one-quarter inch shorter than the distance from top to bottom-bar. It would be better to have them touch both top and bottom-bar, but it would be more difficult to put them in. They are put in something like two inches apart, the two outer ones within half an inch to an inch of the end-bars. The splints are put in a dish of hot wax and left there till all frothing and bubbling ceases, and then they are lifted, one by one, by a pair of pincers, laid upon the foundation, which must be properly supported upon a board, and an assistant presses each splint into the foundation by means of the edge of a little board kept constantly wet. If put in while too hot, there will not be a good coating of wax on the splints. The foundation enters the groove in the top-bar and goes down through the bottom-bar, which is in two parts, the lower edge of the foundation being squeezed between the two parts. The advantage is that the comb is built down to the bottom-bar. You may like better, however, a plain bottom-bar, all in one piece. If, however, such a frame of foundation be given at a time when they are gathering nothing, the bees will gnaw a passage over the bottom-bar.

Q. Where you use foundation splints and split bottom-bars, what kind of foundation do you use—medium or light brood?

A. Medium gives good results, but light brood might be just as good with two or three more splints to the sheet.

Q. (a) On page 393, of "Langstroth on the Honeybee," you advocate the use of wooden splints to support wide strips of foundation. Do you use these splints in extracting frames as well as in frames for chunk honey?

(b) Do you use splints opposite each other on the foundation, or do you use them on one side only?

A. (a) I would use splints in extracting combs, but on no account in chunk honey, unless the honey were afterward to be cut up on the lines of the splints and the splints taken out.

(b) On one side only.

Stimulation.—Q. What is the safest and best plan to pursue to stimulate brood-rearing in weak colonies in the spring, and how long before the honey-flow should one commence? (Wisconsin.)

A. In your locality probably the safest and best thing is to see that the bees have abundance of provisions, and let them entirely alone, for more harm than good may be done by frequent feeding in catchy weather. But in localities where there is nice, warm weather for bees to fly, and nothing to get for a week or more, then it may be a good thing to feed a pound or so every other day.

Q. Would uncapping a little honey every day be as good for

stimulative feeding as syrup made from granulated sugar? There is plenty of honey in the hive, and I want the bees to build up so that they will be booming when the honey-flow comes on.

A. Yes, and very likely you may save yourself much trouble. Every two or three days may do as well as every day. If the queen is laying all the eggs the bees can cover, it is hardly worth while to take even that trouble, for stimulating can hardly help, unless it be that it gets up extra heat in the hive.

Stingless Bees.—Q. Is it true that stingless bees have been produced?

A. There are stingless bees in South America, as has been long known, but they are not of value commercially.

Stings (See Beestings.)

Strawberries.—Q. Do bees gather honey from strawberry blossoms?

A. I don't think bees get much honey from strawberries.

Sugar.—Q. I want to know if all sugar sold for granulated sugar is cane sugar or will answer for bee-feed. How can I tell cane sugar from beet sugar? How can I procure cane sugar in rural sections like the hill towns of New England?

A. I think that only a small part of the granulated sugar is made from cane sugar, and although I have tried very hard to learn some way in which cane sugar could be told from beet sugar, I am still in ignorance on that point. The British Bee Journal stoutly insists that beet sugar should not be fed to bees, but authorities on this side the water insist just as strongly that there is no possible difference between beet and cane sugar when it is made into granulated sugar. Certain it is that thousands of pounds of granulated sugar made from beets have been fed with good results.

Sugar Candy.—Q. How do you make queen candy?

A. Take a small amount of extracted honey warmed and work into it enough powdered sugar to make a stiff dough. Let it stand a day or longer, and if it becomes thin, work in more sugar.

Q. Do you think soft sugar is as good to stimulate brood-rearing as syrup? Is it as good for winter stores? Is soft brown sugar all right for bees?

A. I should think there would be little difference between soft sugar and syrup. But neither of them is as good as honey for brood-rearing. Brown sugar is good for bees at any time when they are flying, if they will take it; but syrup of granulated sugar is better for winter.

Sugar Syrup.—Q. How thin a sugar syrup may be fed to bees without danger of spoiling after taking into the hive?

A. Early in the season, when the bees are flying daily, it will do no harm to feed them syrup just as thin as they will take it, say one part sugar to ten of water. And the same is true until fairly late in the season. As the weather begins to be cool toward fall, the syrup must be given thicker and thicker, lest the bees do not have time to evaporate it sufficiently, and as late as November it will not be well to feed a thinner syrup than two parts sugar to one of water, and two and one-half of sugar to one of water is still better.

Q. In making syrup two to one in boiling water, after it is thoroughly mixed, is it necessary to put it on the stove to let it come to a boil?

A. All that's needed is to dissolve the sugar, even if in cold water.

Q. How would you prepare sugar syrup to feed in cellar, when it is absolutely necessary to feed in mid-winter to preserve from starvation?

A. Just a plain syrup, two pints or pounds of water to five pints or pounds of sugar. Stir the sugar slowly into the hot water, and be very careful not to scorch it. But you might do better to make a plain candy and lay over the frames.

Q. I have half a barrel of common syrup that is of no use to me. Would you let the bees have it in the spring? If so, how, and what time?

A. Yes, there is generally a time in the spring when bees can fly every day but get little or no nectar. At such a time it will be well to feed such syrup. Either feed in the hive with a Miller or other feeder, or if you are situated where neighbor bees will not get the lion's share, you can feed in the open air in shallow dishes with cork chips on top, or some other arrangement to keep the bees from drowning. Open-air feeding may be a little the best for the bees—more like working in the field.

Q. Will you please inform me what kind of acid is used in syrup to keep it from granulating? And how much of the acid is used to a gallon of syrup?

A. An even teaspoonful of tartaric acid for every 20 pounds of sugar is stirred into the syrup about the time the sugar is dissolved. The acid is first dissolved in a little water.

Sumac.—Q. Does sumac yield honey? I removed some honey this season that had a greenish tinge. The comb fairly melted

in the mouth. It is capped white. Could this have been sumac?
The bees worked on it steadily for a week.

A. Sumac is a fine honey-plant in some places. It is very light
amber, usually. Your honey was probably sumac mixed with
other honeys. Sweet clover honey has a greenish tinge.

Sunflower.—Q. Is there any honey in sunflowers, and, if so,
could the common black bees get it?

A. Yes to both questions.

Supers, Examining.—Q. How often should supers be looked
after, or rather, examined, as to how far they are filled?

A. Every ten days is not far out of the way, generally, only
make sure that they are never crowded so as to lack storing-
room.

Supers, Exchanging.—Q. If I take the supers from the parent
colony and put them on the new swarm, will the bees that are in
the field keep on working in the super?

A. Yes, there is danger of the queen going up if the supers are
put on immediately after the swarm is hived, unless a queen-
excluder is used. In the absence of an excluder, do not put the
supers on for about two days, and in that time the queen will have
made a start below and will not be likely to go into the super.

Supers, Failure to Work in.—Q. I have five colonies of bees and
they are doing well, as far as I can see, but they are not at work
in the supers. What is the cause? What is the remedy, if any?

A. There may be several reasons why bees do not work in supers. There may not be a sufficient flow to supply more than their
daily needs. The colony may not be strong enough, and you must
wait until it builds up stronger. The brood-chamber may not yet
be filled, and the first care of the bees is to fill all vacant room
below before storing in the super. Sometimes, however, the bees
are slow about making a start in the supers, when they seem
strong enough, with a good flow, and the brood-chamber filled.
In that case you must put a bait in the super to bait the bees into
it. Just how you will do that depends upon the character of your
supers. If extracting-supers, then you can likely put into the
super a frame of brood for a few days, or until the bees begin
work in the adjoining frames. If you have sections in supers,
then put in the center of the super a section that is partly built
out, either empty or containing some honey. If you can do no
better, you can cut out of one of the brood-combs a piece of brood
or honey and put it in a central section. If that will not start them

to work you may know that they are not strong enough to store in the super, or that there is not enough for them to store. Perhaps the ventilation is not good, and the supers are uncomfortable.

Supers, Kind to Use.—Q. Which section super do you prefer, and would you advise plain or beeway sections, and which kind of separator?

A. After a great deal of experience with different supers, I prefer the "T" super. A considerable experience with different kinds of sections makes me prefer the beeway, 4¼x4¼x1⅞. In handling the plain sections, one must be more careful lest the fingers be thrust in them, and more careful lest they tumble over. A loose, plain wood separator serves well, is inexpensive and easy to clean.

Q. What objections, if any, do you have to the fence and Ideal super?

A. Without mentioning any other objection, the fences are more troublesome to clean, and so are the plain sections that go with them, for it is easier to mar the honey in a plain section, and it topples over more easily when standing.

Q. Which is the better for extracting, the full depth or the shallow super?

A. Except for the inconvenience that they cannot be used interchangeably in the brood-chamber, the shallow frames are considered better for extracting. The queen is less likely to go up into them, and their shallowness makes them easier to clean.

Supers, Number for Each Hive.—Q. Would two supers be enough to get for each hive?

A. You can get along, after a fashion, with only one super, but it is very poor economy to scrimp in the matter of supers. If you mean extracting-supers of the same size as the brood-chamber, three would be better than two. For sections, I would not like to start in the season with less than six supers of 24 sections each for each colony.

Supers, Putting On.—Q. Is there any danger of giving a super too soon?

A. Yes, if you should give a super a month before the harvest it would be making the bees keep up the heat unnecessarily in just so much more room.

Q. If I put on the supers before the bees swarm, will that keep them from swarming?

A. Sometimes it will; generally it will not. Giving plenty of room is one of the things that help to prevent swarming, but it is only a help, and not a reliable preventive.

Q. When is the best time to put supers on hives? Do bees necessarily accumulate on the outside of the hive before swarming?

A. The best time to put on supers depends a little upon what you may desire. If you are anxious for increase through natural swarming, it may be best to delay putting on supers till after the harvest is under way, for crowding the brood-chamber with honey will have the effect of starting the bees into the notion of swarming. Indeed, it would make a more sure thing of the swarming if no super should be given until after the bees have actually swarmed. Generally, however, the desire is for honey rather than swarms. In that case the super should be given before there is any danger of crowding the brood-chamber with honey. A little too soon is better than a little too late. One way is to watch the flowers from which the harvest is expected, and put on supers as soon as they appear in quantity. In your region white clover is probably the thing for you to watch. Another way is to watch the condition of the brood-chamber, and put on supers when the brood-combs begin to be crowded with honey. The old rule was to give supers when white wax begins to be plastered on the upper parts of the comb; a good rule in most cases, but for those who prefer not to have any swarming (even though it may be a rare thing for the bees to respect their wishes) it is better to have supers on before the bees get so far as to secrete this extra wax.

Q. When a super is nearly full of honey, is it best to put another super on top and let the bees get well to working before "under-supering?"

A. When a super is something like half filled and the prospect is good for a continuous yield, put a new super under. You may at the same time put an empty super on top, ready to be put next to the hive at the next shift.

Q. In the early part of the honey-flow, in putting on extra supers, do you put them underneath those already on top?

A. The second super is put under the first, and at the same time another empty is put on top. This last serves as a safety valve in case the bees should need more room. There is another important advantage. With the best care it will sometimes happen that the upper starter will not be fastened securely its entire length, although this would not be noticed in ordinary handling. If such a section be put next the hive under another super, the bees will cluster upon it and drag it down. If it be put on top the bees will very gradually occupy that super, and will fasten the starter securely before any special weight be put upon it. In most cases the top super will not have much done in it, but will be ready to be put down as the lowest one, and a fresh empty super will be put on top. When the flow is on the wane some care must be taken not to have too many unfinished sections, and then the empty super is not put below, but if the bees need more room they can work up into the super on top.

Q. When there is a good honey-flow, and two supers full of honey, would it be best to take the two supers off and put on the third on top of the two, so as to give the honey a better chance to ripen?

A. With a good flow on, it will probably never happen that it will be good practice to take off the two supers that are on, and leave the colony with one empty super. For the bees should always have at least plenty, if not abundance, of room, and so a third super should always be given before the first two are ready to be taken off. In my own apiary, a good flow being on, a super is not often taken off before three or four supers are on, and in a few cases there may be as many as seven or eight on. When the first two are pretty well filled, a third super is given below them, and likely enough another on top. All this referring to a bee-keeper running for section honey. With extracted honey all may be left on until the close of each particular flow, if not to the close of the entire season, or the honey may be extracted whenever it is ripe. The third super is generally given below, a queen-excluder being used. But E. D. Townsend, a very successful bee-keeper, gives the empty super above, dispensing with the excluder. He says the combs filled with honey act as an excluder to keep the queen from going up into the empty super.

Q. How high do you tier up? I am using the Townsend way by putting an extracted comb on each side, and sections in the center, and on some hives I use shallow extracting-frames filled with comb. I find these were one-half to three-fourths filled with honey by June 16, and have put supers filled with sections under the partly-filled ones.

A. In a very poor season there will be no tiering up. In a good season, after the season has fairly advanced, there will be three or four supers on each hive, and from that up to seven or eight. But in the latter case the top and the bottom super will likely be empty, or nearly so.

Q. With a full depth extracting-super, would it be any advantage to put the colony with the queen above the super with a honey-board between? Would the bees store honey in it at all? I thought perhaps the bees, having to pass through the super to get to the brood-chamber, they might store some honey earlier than otherwise.

A. Bees prefer to store honey above their brood, and with room above, you could hardly expect them to store below. Yet in a strong flow I have had them store in a story below. But they will not store below so soon as they will above.

Supers, Removing.—Q. The super is now nearly full. Is it advisable to take it off?

A. It is well to take away sections as often as a complete superful is ready, although it is hardly best to wait until the corner sections are all sealed, for if you do so the central sections will have their cappings darkened. The unfinished sections may be assembled from different supers into one super and returned to the bees to be finished.

Q. I run my bees for extracted honey. How can I free the supers of bees without having to brush every comb? I do not care to use the Porter bee-escape board if there is any other way.

A. You could use some other escape, as the Miller tent-escape. It consists of a robber-cloth with a cone of wire-cloth centrally located. The La Reese escape is favored by some. The Miller escape is put on top of a pile of supers after they are taken from the hives.

Q. Is it proper to take off the supers in the fall of the year? (Arkansas.)

A. In your latitude (36 degrees) it will probably do no harm to leave extracting-supers on the hive over winter. But it will not do to leave section-supers on the hive over winter in any climate, because the comb in the sections will be spoiled. Neither should the sections be left on until fall, unless the honey-flow continues until then. Just as soon as the bees stop storing in the sections they should be taken off.

Supers, Space Below.—Q. How much space should there be between the brood-frames and supers, or board, if left on?

A. About one-fourth inch; less rather than more.

Supers, "T."—Q. What is a "T" super? Looking up the catalogs I find nothing except "T" tins in this line.

A. A 'T" super is a plain box without top or bottom, one-quarter inch deeper than the height of the sections it is to contain. On the bottom, at each end, is a plain strip of tin to support one end of the sections in the end rows, and at the proper places staples are driven into the bottom and then bent so as to support the "T" tins inside. On page 19 of "Forty Years Among the Bees" is a picture of a "T" super, which is reproduced here. I'm sorry

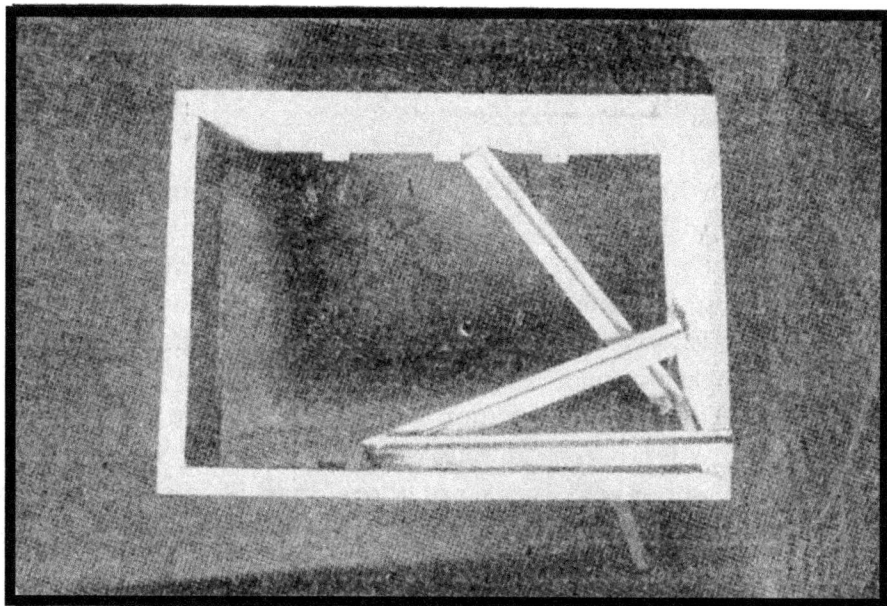

FIG. 24.—A T-Super.

to say it doesn't show as plainly as it might what a "T" super is. The three "T" tins are shown loose, and you will see at the bottom of the super the supports for them, which are here squares of sheet-iron nailed on. The bent staples are later, and perhaps a little better.

It seems a very strange inconsistency that allows "T" tins to be listed in a catalogue and not the "T" super, for without the "T" super one will have no use for a "T" tin. For some reason, no manufacturer pushes "T" supers, and yet there are not a few who produce section-honey on a large scale who will have no others. As for myself, I have tried about all the surplus arrangements for section-honey that have been put on the market, some of them on a pretty large scale, and as yet have found nothing else to equal the "T" super. I have seen it condemned, but when I learned how it was used, without taking advantage of its best

features, I don't wonder at its being condemned. I have no personal interest in the affair; it is no invention of mine, but it is
my deliberate conviction that at present there is no better super
in existence that the "T" super.

Q. Just how long ought the "T" super to be made inside? .

A. Mine are 17⅜ long inside. I don't know whether any
other length would do better.

Q. I read that you use the "T" supers. I have a few regular
supers on hand, but figuring how much furniture it takes, and the
trouble to keep them clean, I thought perhaps this was your reason for using the super you do.

A. My reason for using the "T" super is that I think I can
produce section honey of fine quality with less labor and expense
than with any other kind I have tried, and I have tried many
kinds. I think very few who have tried the "T" super probably
have given it up. Some who have condemned it have never used
it properly. I know of no super that allows the same number of
sections in more compact form. When four supers are on a hive
—in a good season it is a common thing to have four to six supers
on a hive—the distance from the top of the lower section to the
bottom of the upper section is not more than 9½ inches. It does
not seem possible to invent any super that will allow the sections
to be in less space, for no room is taken up with bars under or
over the sections. In most other supers there is a bottom-bar under the sections, and in some a top-bar as well. In the latter case
even if top and bottom-bars be only one-quarter inch thick, the
distance between upper and lower sections, instead of being 9½
will be 11 inches. But a bottom-bar one-quarter inch thick is
likely to sag, and even one-half inch may sag through warping
The "T" super has the advantage that the "T" tins are entirely
rigid, with no sagging whatever. I might go on and tell hov
easy it is to fill the super with sections, and how easy to clear
the sections. All these things, when properly done, set the "T
super at the head, in my judgment, as the best super for producin
comb honey. Along with this is the fact that it costs less.

Q. Will the "T" supers fit the standard hives? If not, I coul
not use them.

A. A "T" super will fit any hive that is flat on top, which i
pretty much the same as saying it will fit on any hive. My super
are rather short to fit my hives. I count that an advantag
Sometimes I want the super to be shoved just a trifle forward t

allow a quarter-inch space for ventilation at the back end. When I don't want that ventilation I tack on the super at the back end a strip as long as the width of the hive, or the super, and about seven-eighths by one-half. That makes the super long enough so it covers entirely the opening at the top of the hive. The super being made just as wide as the hive, of course it will be wider for a 10-frame hive than for an 8-frame hive.

Q. Do you use springs with the "T" super? If so, how many, what kind, and how?

A. I use a single spring in each super, crowded in between the follower and the side of the super. It is the common super-spring sold by supply dealers, in shape something like the elliptic of a buggy.

Q. I think you spoke of driving the bent staples in level with the wood. If I understand it rightly, the T-tins are supported just the thickness of the staples above the bottom of the super. Am I right?

A. The idea is to get the bottoms of the sections as nearly as possible on a level with the bottom of the super, but in actual practice that will vary, for in bending over and driving down the staple it will be sometimes embedded in the wood.

Q. Are the thin strips of wood better than an extra set of tins at top of the sections? I was under the impression that the tins were better.

A. The wood strips leave propolis only at the corners of the sections, while "T" tins on top would invite lines of propolis at some distance from the corners. The wood is probably easier to put in place, and it holds the sections square, while a "T" tin on top would allow a little variation. Yet these differences do not amount to so very much.

Super, Turned Over.—Q. Why wouldn't it be a good plan to have the super so that you could turn it over? Wouldn't we get better filled sections?

A. Supers have been made to use in that way, but have never come into general use; perhaps because the advantage did not pay for the extra trouble.

Supersedure.—Q. Is there any way to tell a supersedure queen-cell from a swarming cell, during the swarming season?

A. There is no difference between a supersedure-cell and a swarming-cell, either in appearance or any other way. It may

happen that the bees start to supersede a queen without swarming, and then conditions for swarming turn so favorable that they
swarm. Again, bees may prepare to swarm, when conditions for
swarming turn so unfavorable that they give up swarming. In
that case they may simply destroy the cells and allow the old
queen to continue, or they may supersede the old queen. If, during the swarming time, you find queen-cells, you may be almost
sure it means swarming. If cells are found somewhat out of the
time when most colonies are swarming, you can only make a
guess in the case. If the number of cells is small—not more than
three or four—and especially if the queen is old, it is likely to
mean superseding. For swarming, a larger number of cells will
generally be found.

Q. Will a good colony supersede its worn-out queen, or is it
not best to introduce a new queen at least every three years?

A. Opinions are divided. It is possible that locality may have
something to do in the case, as it has in so many other cases. In
this locality it is as well to leave the matter to the bees, generally,
although it pays any time to supplant a poor queen with a good
one, even if the poor one is only a month old.

Q. In Doolittle's "Queen-Rearing," page 111, he says: "To
supersede a queen, hatch a young queen in an upper story over a
zinc excluder, and after she is hatched remove the excluder, and
your old queen is superseded." Will the plan work invariably?

A. No; and I do not think Mr. Doolittle claims invariable success. Remember that in the natural course of events every queen
is superseded by the bees, and that such a superseding usually occurs somewhere in the neighborhood of the close of the harvest.
Now, when any colony has a queen that it is about to supersede,
if you will get in a little ahead by having over the excluder a virgin before one has been reared below, you may be practically
certain of success. If you do the same thing earlier in the season,
especially where a vigorous queen is doing duty below, you may
expect failure.

Q. Yesterday (Feb. 17) was the warmest day we have had
here this winter, 60 degrees in the shade for the greatest part of
the day. I took my bees (ten colonies) out of the cellar for a
flight, and found on looking them over that one colony had a
patch of drone-brood about 3 inches in diameter, partly capped
over on both sides of one comb. I found some worker-brood in
the rest of the hives, but this one had none. I found the queen,
but she looked more like a virgin than a fertile queen. Do bees
supersede their queen in winter?

A. In the proper sense of the word I doubt that a queen is ever superseded in winter. If a queen is lost, they may try to replace her almost any time. Your queen is a drone-layer, and so worthless.

Q. What causes supersedure when everything apparently looks in good condition. September 2 I had a swarm go out, and upon examination of the hive I found that they had superseded their queen (which was of this year's stock), and there were also four other virgins in the hive. I knew it was too late for a profitable swarm, so I pinched the heads off of all but one queen, destroyed all remaining cells, and then put the swarm back in the same hive. Was this right? Everything is going along smoothly at this date (Sept. 5.), and the new queen is laying.

A. You ask what causes supersedure when everything apparently looks all right. That "apparently" is probably the answer. A queen may be in some way at fault, whether a few days or a few years old, and you may see nothing wrong, but some way the bees know about it. It is not entirely clear, however, from what you say, whether this was not a case of regular swarming, rather than supersedure. In any case, you did well to do as you did.

Q. How is it that bees neglect to supersede their old queen when there are drones to mate with the young queen, as this has happened to me several times late in the fall?

A. If I understand correctly, you have had queens die in late fall or early spring when there were no drones, and your question is why they didn't supersede them earlier, when plenty of drones were on hand? I don't know. It is possible that some accident may befall a queen, and of course the bees could not foresee this. It would seem that bees recognize the trouble when a queen begins to fail, and supersede her; and it is possible to conceive a case in which there was no sign of failure while drones were still present, but an unusually rapid failure after they were gone. The fortunate thing is that such cases are rare; nearly always a queen is superseded with abundance of drones present.

Swapping Combs.—Q. Would what is called "swapping combs," i. e., taking a frame of comb or foundation from the surplus-box, and exchanging it for a frame of brood, tend to get the bees to work in the super and also tend to prevent swarming?

A. It would tend to start the bees to work in the super, but would not do much to prevent swarming.

Swarm-Box.—Q. How can I use a swarm-box as mentioned in "First Lessons in Beekeeping?"

A. A swarm-box being lighter than a hive, instead of carrying the hive to where a swarm is, the box may be taken there, and when the swarm is in the box it can be carried to the hive, laid upon its side with the open part of the box toward the entrance of the hive, so the bees can run from the box into the hive. If they are too slow about it they can be dumped on the ground in front of the hive by jarring the box on the ground.

Q. What are the best noticeable signs just before swarming?

A. The most reliable sign that bees meditate swarming is the finding of a number of queen-cells in the hive.. You may, however, judge a little from outside appearances, if you find a colony ceasing work and loafing when other colonies keep on at work; and when bees return from the field laden with pollen and join the outside cluster without going inside to unload.

Q. Last spring I bought a colony of bees and was very anxious to have them swarm. The first swarm issued July 13. July 20 the mother colony swarmed again. This swarm covered six frames. On July 24 the third swarm issued from the parent colony. A week later I opened the parent colony and found that the bees had done nothing in the super. The body of the hive was full of honey, and I found three queen-cells. Two of these I destroyed. The cap of the third seemed loose, and soon the queen crawled out, at least I thought she was the queen, though she looked like any other bee. Do you suppose I have left the colony queenless?

Swarm No. 1 has made lots of honey, while the other two swarms and the parent colony have made nothing. Had I better unite these, and how, or would it be better to give them frames of honey from the other hive? Should I get new queens for the two latter swarms and for the original colony? Should I go over the combs every ten days and cut out queen-cells?

A. There is nothing unusual in the program your bees have followed. The mother colony having sent out three swarms, has not bees enough left to do anything in the super, and all the bees are crowded into the brood-chamber. Neither are the second and third swarms strong enough to do much, the first swarm being the only one strong enough to do super work. When a colony prepares for swarming, it starts quite a number of queen-cells, and you found what were left after the last swarm issued. It is not likely that your cutting out those last cells made any difference about swarming, for it is a rare thing for the fourth swarm to issue. You may or may not have made the colony queenless by cutting out the cells. You say the bee that came out of the cell looked just like any other bee. It is quite possible that it was a worker. Sometimes a worker crawls into a queen-cell after the queen has left it, although the capping of the cell looks as if the queen has not yet emerged. If that was the case, then the queen was left in the hive and the colony is all right. You cannot be certain about the queen by the carrying of pollen. If you do not find eggs in the hive about 10 days after the last swarm issued, or at least in two weeks, you may decide the colony is queenless, in which case you will give it a queen, unless you prefer to unite it with the weakest afterswarm. The chances are that both afterswarms have queens all right. The likelihood is that they will build up without any help from the first swarm, which can be left undisturbed at its work of gathering honey. Of course, if the bees do not gather enough for winter you will have to feed. It is not likely you will have any difficulty in telling a queen when you see one, by its greater size, especially length. No need to go over your hives for queen-cells now, after swarming is over.

Swarm Prevention.—Q. Do you like to destroy all queen-cells but one, or clip the queen's wings for the prevention of swarming?

A. To prevent a prime swarm, neither one will answer. Destroying not merely all but one, but **all** cells, will generally delay swarming, and sometimes prevent it, but too often the bees will swarm in spite of cell killing. Clipping the queen doesn't have the slightest effect in preventing swarming. All it does is to prevent the queen from flying with the swarm, and when the bees find that the queen is not with them they return to the hive. But if the beekeeper does not interfere, the bees will swarm just as soon as a young queen is reared.

Q. Will a young queen swarm out after she commences to lay? I can't remember having had one do so where I knew the queen to be a young one. Some writers claim a queen never lays drone-eggs the first season, and I never found queen-cells started in a normal colony without more or less drone-brood in evidence.

A. The answer to your question must be a little mixed. If you allow a young queen to be reared in a hive, you may count on no swarming before the next season. If you introduce a young queen, it depends. If the colony is in no humor for swarming at the time the queen is introduced, then no swarming. If in the swarming humor already, they'll swarm in spite of the tender youth of the queen. I once had a queen swarm in less than a week after being introduced, and she had been laying only about a week. If I had kept the colony queenless until swarming had been given up, and then introduced the young queen, it would have been all right.

Q. There is usually a fairly good fall flow here of aster, goldenrod and buckwheat, and I would like to know if caging the queen in June or July to prevent swarming would be practiced at a loss in regard to fall honey. Would not the removal of the queen for ten days during June result in the loss of about 20,000 bees, figuring 2,000 eggs a day, that would be ready for a fall flow August 15?

A. You are probably overestimating the number of eggs laid daily. If we allow three-fourths of the frame to be occupied with brood, a queen laying 2,000 eggs daily would keep eight frames occupied. I don't think many queens do that when the season is so far along. Whatever is the right figure, it will be just so much loss in your honey crop. But the loss would likely be greater still if swarming were allowed.

Q. How would you prevent swarming? I have 15 colonies and they do nothing but swarm. I give them supers with starters, and they will go up and fill two or three sections and then swarm. One of my colonies swarmed four times in a week and a half. What would you do to stop them from swarming?

A. It is not an easy thing to prevent a first or prime swarm. Perhaps what will suit you as well as any way is to allow the first swarm to issue, and then prevent afterswarms in the following way: Set the swarm on the stand of the mother colony, putting the old hive close up beside it, both hives facing in the same direction. A week later move the old hive to some new place six feet or more away. That's all; the bees will do the rest, and you are not likely to have any further swarming from a colony thus treated.

Q. When bees swarm you say hive the swarm, place it in the place of the old hive close by, and a week later move the old hive away to its future place. If you follow this plan, will the old colony store any surplus? If not, will the swarm make up for it?

A. Unless the season is very good there will be little or nothing stored by the mother colony, but the swarm will store more than both would have stored if the swarm had been put on a new stand and the mother colony left on the old stand.

Q. Is the following method all right to prevent increase: Let the swarm issue, kill the queen and send the swarm back; wait seven days, then cut out all queen-cells but one. I tried this

method with one colony, and it was quite successful. Will it always be successful?

A. The plan is good and will usually be successful; but
sometimes you may miss a cell, and sometimes the only cell you
leave may be bad.

Q. If a young queen is given to a colony in the spring, will
swarming be retarded, and if so, to what extent? That is, how
much more crowding will they stand, or the reverse?

A. If a queen that has been laying only a few days be given
at the beginning of the swarming season, and if the colony has
not yet made preparations for swarming, there is very little
chance of swarming that season. The same is true to a greater
or less extent if the young queen be given earlier. I am not sure
about the retarding, but the chances for swarming are greatly
lessened by the giving of a young queen. When you ask me to
tell just how much crowding they will stand, you're crowding me
in too tight a corner. Fact is, I don't know. I think something
depends upon the queen, and perhaps still more on the bees. With
some bees, a vigorous young queen could probably not be forced
to swarm by any amount of crowding, provided the queen were
not given too early, and from that it will shade all the way down
to where allowing only room for 25 pounds of honey might induce
swarming.

Q. What per cent of swarming do you have, in spite of all
your preventatives?

A. I count the prevention of swarming an unsolved problem.
At a rough guess I should say that there may be from 5 to 10
per cent of the colonies actually swarm. But if they do swarm,
nc swarm is ever hived as a separate affair, but obliged to remain
in its old colony, for one of the important points in securing
gcod yields is to keep from dividing the forces.

Q. I want to tell you of a colony I have which swarmed on
May 30, June 30 and July 30. Each time it was treated by the
put-up plan. I want to ask what to do with this queen? She has
swarmed out three times so far, and has made twice as much surplus in sections as any other colony I have. She is a nice, large
queen, very prolific; but I don't like this thing of swarming.
Would you breed from such a queen? I like her because she has
such nice workers, busy all the time, and, as I stated, made me
more surplus by far than any other colony.

A. Generally, after a colony has been treated by the "put-up"
plan, there will be no more swarming for the season, but you can

never be certain of it. Yet it is a rare thing that a colony swarms a third time, as in your case. Yet I should not be much afraid to breed from such a queen if the colony greatly exceeds other colonies in storing. (See "Put-up Plan.")

Q. To prevent swarming, I will shove the pile of supers back so as to make an opening of one-half inch for ventilation along the front. Will this prevent swarming and affect the storing of honey?

A. Ventilating in that way is a help against swarming, although, of course, it will not prevent it. I have practiced it much, and never knew any harm to come from it in the way of chilling bees. The only harm is that sections next to the opening are delayed in being finished, but not always. Instead of being shoved back, I shove the lower section-super forward. I have used the plan with extracting-supers, and "stuttered" them; that is, I shoved the lower super forward, the next back, the next forward, and so on.

Q. If the queen is given plenty of room, will swarming be prevented, even though the hive be crowded with bees?

A. It would certainly decrease the tendency to swarm, just as increasing the queen's room for eggs always does; whether it could be relied on in all cases as an entire preventative is hard to say without trying. I should rather expect it would, so long as fresh room for the queen is constantly given, and even when the flow comes and the lower hive is given above as an extracting-super, there ought to be little inclination to swarming, as in the case of the Demaree plan.

Q. I kept my bees from swarming two years ago by placing the brood over the queen with an excluder between. When buckwheat came on, I had my hives chock-full of bees. I also had several swarms in September, and that's rather late for Northern New York. I'd like to know how to stop them at that time.

A. Yes, that's the plan given by G. W. Demaree, a Kentuckian who was prominent in the ranks some years ago. The plan is good and the pity of it is that it will work only for extracted, not comb. The brood-combs being put above become extracting-combs. To prevent swarming in September (which is not usual, I think, but may come where there is a fall flow), it might work to try the same plan over again; extract the frames in an upper story, put them in the lower story with the queen, the brood above, excluder between. If this is done just as buckwheat begins, it seems it ought to work as well as earlier.

Q. I would like to ask some questions about the Demaree plan. You say just before swarming, put all the brood but one frame in a second story over an excluder, leaving the queen below with one frame of brood and empty combs or frames filled with foundation.

Do you cut out all queen-cells at this time if there are any?

Is it necessary to examine each colony in about ten days to remove queen-cells afterwards?

A. Yes, to your first question.

As to the second, generally it ought not to be. The idea is that the bees are in the same condition as if they had swarmed naturally, of course, it sometimes happens that when a natural swarm is hived it throws off a swarm the same season, but that is exceptional. Some have reported that they never have a colony swarm that has been treated by the Demaree plan, while it fails with others.

Q. I have four colonies in a house apiary I want to prevent from swarming. Would it do to add a hive-body with wired foundation below, as soon as the queen needs the room, then about three weeks before clover, or about May 20, put the queen below, then an excluder, then a super of shallow extracting-frames, and over all the old hive-body, with brood, and about June 10 remove the old hive-body from the top and put comb-honey super between the extracting-super and the excluder? Provided I give ample room, would that be likely to prevent swarming?

A. Unless your bees are unusually "forward looking" they may be behind time on the program you are laying out for them. You say add a hive below "as soon as the queen needs the room," and evidently expect her to need it so early that she will have the brood-nest extended into the lower story by May 20. Maybe she will. At any rate it will do no particular harm to have the empty story below. Suppose there is nothing doing below, and May 20 you put the queen on the foundation under the excluder. In too many cases the queen will swarm out, unless you put something in the way of bait below. At any rate, I've had them swarm out. Suppose, however, that the brood-nest is started below, or if not, that you give a frame of brood. The bees will go to work all right (you must look out for cells in the old brood above); they will fill up the lower story, and then swarm. Not always, but I should expect it to happen a good many times. . They will not be so certain to swarm as if you had let them alone, nor will they swarm so soon. But you have operated so early that you may expect more swarming than you want. The later in the season you give the queen that empty story below, the more certain you will be to have no swarming. Put it off just as long as you can without having the bees actually swarm. If you wait until cells are started, and then operate, destroying the cells, you may feel pretty easy about swarming. Some report it a perfect preventive. You propose to put a comb-honey super under an extracting-super. That will be all right if the extracting-combs are nice and white. If black from brood-rearing your sections may be blackened.

Q. Will you please explain Mr. Allen's system for swarm prevention that he gave to the readers of the Bee Journal several years ago? If it really has any merit, will you kindly reproduce it in the Journal?

A. If you will turn to page 94 of the American Bee Journal for 1910, you will find the plan given by A. C. Allen, as follows:

"When the honey-flow is well started, I go to each strong colony, regardless of whether the bees desire to swarm or not, and remove it from its stand, putting in its place a hive filled with empty combs, less one of the center ones. Next, a comb containing a patch of unsealed brood about as large as the hand, is selected from the colony and placed in the vacant place in the new hive; a queen-excluder is put on this lower story, and above this a super of empty combs, this one having an escape hole for drones; and on top of all, an empty super. A cloth is then nicely placed in front of this new hive, on which the bees and queen are shaken from the combs of the parent hive, and the third story is filled with the combs of sealed brood and brood too old to produce queens, and allowed to remain there and hatch, returning to the working force."

This is really the Demaree plan, which was given to the public many years ago, by G. W. Demaree, a prominent Kentucky bee-keeper at that time. Mr. Allen has varied it by putting a frame with some brood in the lower story, whereas I think Mr. Demaree had only empty combs, or combs with starters in the lower story. Mr. Allen's variation is of value, for I think there were cases reported in which the bees swarmed out with no brood in the lower story. Mr. Demaree put all the brood in the second story, while Mr. Allen puts it in the third. I don't know which is better.

Mr. Allen says "the third story is filled with the combs of sealed brood and brood too old to produce queens." I hardly understand that, for he says nothing about putting brood elsewhere, and generally most of the combs would have at least some very young brood.

The plan is a good one for extracted honey, but not available for comb.

Q. The other day a book came to me from a friend in Missouri. It was Dr. Jones' book on how to prevent swarming. What do you think of the plan?

A. I think that shaving off the heads of all sealed brood will be likely to prevent swarming; but I don't suppose many would want to use such a plan.

Q. In the American Bee Journal for November, 1914, page 385, the plan given by J. E. Hand on swarm control and the increase problem looks possible, and I would like to have your opinion of it. I note he uses 16-frame hives. Do you think the plan will work with the 10-frame hive? He says take six frames from each colony at the close of the basswood harvest and give them to the nuclei. But that will not be necessary, as I can build a 2-frame nucleus into a full colony by winter. Is it necessary to wait until each colony has made preparations to swarm, or can it be done just before the swarming season?

A. As a rule, it is not well to attempt any changes on any plan given, but to follow out exactly instructions. A plan that succeeds with 16 frames might be an utter failure with a less number.

When colonies in general are making preparations for swarming, it will usually be all right to operate at that time upon other colonies that have made no such preparations, provided they are strong; for some colonies make no preparation for swarming throughout the entire season.

Q. In the September issue, 1914, page 310, is an article by C. F. Greening on "Controlling Swarming," which I find most valuable. I wish to ask a few questions concerning it.

(a) It being supposed the colony is a strong one, and of course no queen-excluder being used, will the queen always lay eggs in the super added to the brood-chamber "as soon as it becomes warm" in the spring? In case she does not, this plan is doomed to fail at the very start.

(b) In case she does not come to lay in the super, what shall I do?

(c) Would this plan work with large brood-chambers, such as the Dadant, the Quinby, or the Massie hive, which has a double brood-chamber of a capacity equal to 14 Langstroth frames? With such large hives, how can I make sure the queen will lay eggs in the first super added in the spring?

A. (a) I don't think you can always be sure of the queen

going above to lay, especially if the hive be very large; but the plan is not necessarily doomed on that account.

(b) Take a frame of brood out of the brood-chamber and put it up.

(c) You seem to think that an essential part of the plan is for the queen to go up to lay in the story above. If that be so, then a very large hive would not do. But I hardly think that is essential. If I understand Mr. Greening aright, he wants brood always above, with plenty of room for the bees to store between that and the brood-chamber. The large size of the hive would not interfere with that. Indeed, if I am not mistaken, with the very large hives used by the Dadants they have very little swarming, even without keeping brood above.

Swarms, Cause of.—Q. I have noticed that the colony that keeps its brood-nest free from honey is not apt to swarm. Is this in line with your observations?

A. Certainly we know that when the queen is crowded for room it tends toward swarming. The more honey crowds into or encroaches upon the brood-nest, the less room the queen has for laying, and, it would seem not unreasonable to argue, the greater inclination to swarming.

Swarms, Hiving.—Q. In "Root's A, B, C and X, Y, Z," page 553, 1913 edition, "a frame of unsealed larvæ" is thrust into the bees of a swarm, so that they may crawl upon it. Are unsealed larvæ better than sealed for this purpose?

A. As between sealed and unsealed larvæ, if you have either kind alone, I think sealed would work better than unsealed. When, however, you take a frame having unsealed larvæ, you generally have sealed too.

Q. In hiving a swarm, do you put them in at the top, or the bottom of the hive?

A. It doesn't matter so you get them into the hive, but generally you will find it easier to get them into the hive at the bottom, as bees of a swarm naturally incline to crawl upward.

Q. Is it best to put a swarm on the old stand? Where would you put the old hive?

A. Yes, unless you want afterswarms. The old hive should be put close beside the hive containing the swarm, and a week later moved to any place you like, ten feet or more away.

Q. If a swarm lights on the grass and you do not happen to see the queen, how would you hive them?

A. Set the hive on the ground with the entrance close to the bees, put a few at the entrance, and let the rest follow.

Q. I have just started in beekeeping. I keep my bees in a large attic. Having only one colony, I would like to get another swarm started in swarming time. Can they be made to go into another hive while swarming?

A. Of course a swarm can be hived in any hive, just the same as if the bees were on the ground instead of being in an attic. What you probably mean, however, is to have the bees enter the hive of their own accord. That can be managed, too, just the same as if the bees were on the ground. Have the queen's wing· clipped. Then when the swarm issues catch and cage the queen, move the old hive to a new place and set the new hive where the old hive stood. The swarm, having no queen, will be sure to return within a few minutes, and finding the old hive gone, will enter the new one sitting where it left the old one. As the bees are entering the hive, free the queen and let her run into the hive with them.

Q. In hiving a swarm where only the queen and not quite two frames of bees can be captured, should the division-board be used as in nuclei, and about how many frames of comb or foundation should be given them?

A. A division-board is not needed, but it is better to have a dummy; that is, a board like a division-board, but having a space all around. The hive may be filled at first with combs, or you may give only one comb more than the bees can cover, adding others as they are needed.

Swarms, Issue of.—Q. What time of the day do bees swarm?

A. The prime swarm, which issues with the old queen, generally comes off somewhere between 9 and 12 o'clock. An after-swarm, having a virgin queen, is more irregular, and may issue earlier or later, in some cases as early as 6 in the morning, and as late as 4 or 5 p. m. If a morning is very hot, a swarm may come out early. If the day should be rainy, and clear off rather late in the afternoon, a swarm may come then.

Q. Do bees find a home before they swarm?
A. Often, and perhaps generally.

Q. Are there any indications to tell when the first swarm will issue?
A. Yes; when a colony decides to swarm it starts a number

246   DR. MILLER'S

of queen-cells. About the time the first queen-cell is sealed the swarm issues.

Q. Will a swarm fly out before a queen-cell is sealed?
A. Sometimes; but generally not.

Q. When a swarm issues from a hive, does the queen go out first?
A. No; she may be among the last.

Q. When several swarms issue at once, if the queens are clipped, will all the bees go back to their own hive?
A. As a rule each bee will return to its own hive. In a large apiary, however, it sometimes happens that there will be some mixing, part or all of the bees of a swarm being attracted by the noise of a returning swarm that had previously issued.

Swarm, Kind of Bees in.—Q. When bees swarm, which leave the hive, the young or old, and are they forced out by the remaining bees?
A. Bees of all ages are in the swarm, and they go out without any forcing.

Swarms, Late.—Q. Is a swarm worth hiving after the last of May, and how late are they worth saving?
A. In my locality nearly all swarms issue after the last of May. A good swarm is worth saving, no matter how late it comes.

Swarms Leaving.—Q. How far will a swarm go off to a tree?
A. Nothing definite about it. They may go a few rods or a few miles. They are likely to go to the nearest place where they can find a suitable lodging.

Q. Is there any way to stop a swarm of bees that is passing by you, going to the woods? If so, how?
A. Some have reported success by flashing upon the swarm the reflected rays of the sun by means of a looking-glass. Perhaps the most reliable thing is to throw upon the bees a strong spray of water, since this may hit the queen and make her fall to the ground.

Q. Please tell me how to keep the bees of a swarm from swarming out again. I lost several swarms this summer by their swarming out. They were put in new hives that I bought two years ago.
A. Years ago it was more or less the practice to wash out the hives with this or that under the notion that certain odors made it attractive to the swarm put in it. Nowadays nothing is wanted

but a clean hive. Your hives were, no doubt, all right, but it's a pretty safe guess that the bees were uncomfortable for all that. Either they were too warm or had too little air. Likely both. When you hive a swarm see that it has abundant ventilation. Give it as large an entrance as you can. If practicable, it is a good plan to raise the hive an inch or so from the bottom-board by putting blocks under the corners. Shove the cover forward so as to make an opening of half an inch, or an inch, at the back end. After two or three days you can lessen the ventilation if you think best. The hive should be in a shady, airy place. If you cannot give shade in any other way, cut an armful of long grass, put it on the hive, lay two or three sticks of firewood on it to keep it from blowing away. Some make a practice of giving a frame of brood to the swarm. The bees think that it is such a good start toward housekeeping that they are unwilling to leave it without great provocation.

Some secondary swarms leave because their queen has not yet mated, and they follow her when she goes out for her wedding flight. Nothing will hold such swarms except killing the queen. Then they would return to the parent hive.

Swarms, Moving.—Q. When is the best time to move a swarm after it is hived?

A. Right away after you get the bees of the swarm in the hive. Don't wait to get a few scattering bees in; they can find the swarm where you put it, or else they can go back to the old hive.

Swarm, Prime.—Q. How long after the prime swarm issues forth does the young queen hatch?

A. Ordinarily the first virgin leaves her cell about a week after the issue of the prime swarm. If, however, the swarm be delayed a day or more by bad weather, then the time of her emergence after the swarming will be lessened a day or more. It may also be increased in case the prime swarm issues before the first queen-cell is sealed.

Swarms, Returning.—Q. What is the best manner of returning a swarm to the hive from whence it issued, so as to make it stay, no further increase being desired?

A. It doesn't matter how you return the swarm; it will stay as well for one kind of returning as another. It is the condition of things in the hive that decides whether the swarm will issue again, and it isn't the easiest thing in the world to prevent it. The old-fashioned way was to return the swarm every time it issued, and if you don't mind the amount of work involved in returning it half a dozen times or more, the old way is good. Here's another way you may like better: When the swarm issues, return it and kill the old queen. A week later destroy all queen-cells but one. If you miss no cells there ought to be no more swarming. If you hive the swarm in a box and wait twenty-four hours before returning it to the mother colony, there will be more likelihood of it remaining than if returned at once.

Q. My prime swarms invariably, after I hive them, leave the hives and return to the parent colony. My hives are all new and up to date, and I can't account for this. If you can cast a little light on this subject it will be appreciated.

A. That's just the way my bees do, and it's a good deal better than to have queen, bees and all going off to return no more. The reason my bees do so is because the queens' wings are clipped so they cannot go with the swarms, and when a swarm finds out the queen is not along there is nothing for it to do but to return to the old home. Of course, I don't know anything about it, but as you have things "all new and up to date," my guess would be that you have lately begun beekeeping and have bought colonies with queens whose wings have been clipped. If that isn't the right answer, then I don't know what is the answer. It occasionally happens, where queens are not clipped, that a swarm returns because something has happened to the queen so she cannot fly, but to have it happen "invariably" with whole wings is something beyond me.

Swarms, Second. (See Also Afterswarms.)—Q. After the first swarm issues, how soon can I look for another?

A. A second swarm usually issues about eight days after the first, but the time may be less, and it may be more. The issuing of a prime swarm is sometimes delayed by bad weather, and it may be delayed by the queen failing for some reason to go with the swarm.

Swarms, Sprinkling.—Q. Do you think best to sprinkle bees with water before putting them in the hive when they swarm?

A. It is not a general custom, but if there is fear that the swarm will go off, sprinkling will help to prevent it.

Swarms, Value of.—Q. Will a new swarm gather any surplus honey the first season?

A. Yes; as a general rule the swarm is the one to rely on for a crop, it being put on the old stand after removing the old hive to a new stand.

Q. "If a swarm in July
   Is not worth a fly,
   Can anybody remember
   What they are worth in September?"

A. "A swarm in May is worth a load of hay;
   A swarm in June is worth a silver spoon;
   A swarm in July is not worth a fly."

That jingle must have been made for some locality with which
I have no acquaintance. Taking it, however, at its face value, if
it teaches anything it teaches that the worth of a swarm as the
season advances is a constantly diminishing quantity. In July it
gets down to the zero point, after that it becomes a negative
quantity, by September becoming a great deal less than nothing.
And that might be literally true at the time the doggerel was
composed, when all that was done with a swarm was to dump it
into an empty box or skep and leave it to its own devices. For
the swarm would be worthless, and the mother colony would be
damaged by the exodus. It is possible that in the present instance
there was an exception, and that the flow was so heavy and continued so late that, left to themselves, the swarm might have
built combs and stored enough for winter. In localities where
there is a dearth in July and a second crop in August and September, a September swarm will be better than a July swarm, as it
may fill its hive from the fall blossoms, while the July swarm
would starve before the second crop opened.

Swarms, Where From.—Q. If a swarm comes forth, and you
don't see what hive it comes from, is there any way to tell what
hive it comes from?

A. Take a bunch of bees away from the swarm, dredge them
with flour, and watch to which hive they fly back. Of course, you
may also be able to make a good guess by looking into the hives
and finding one which has a scarcity of bees. If you investigate
the matter promptly you may find in front of the mother hive a
number of very young bees, unable to fly, who have been dragged
out by the rest of their comrades and are trying to get back.

Sweating of Bees.—Q. Do bees sweat if covered too warm?

A. They are more likely to sweat when too cold, if you may
call it sweat. Moisture is coming from the bees all the time, and
if the walls of the hive are cold, the condensed moisture settles
upon them, and may run down and out of the entrance, and this
is sometimes called sweating. The worst of it is when the mois-

ture collects overhead and drops down upon the cluster. Covering up warm helps prevent this.

Q. What makes bees sweat in the cellar in winter? Mine are all wet. I put them into the cellar just as they were in the summer.

A. The moisture from the breath of the bees settles on the cold walls of the hive, just as we say a pitcher sweats when a pitcher of cold water stands in a hot and moist time and the moisture of the air settles on the outside of the pitcher. It is a bad thing to have this moisture settle on the hive-cover, for then the drops fall on the cluster of bees. The matter may be helped by enlarging the entrance, by allowing a little crack at the top for the moisture to escape, or by having some kind of warm packing on top.

Tar Paper.—Q. Is tarred paper injurious to bees and honey?

A. Not in general. If honey were kept for a time directly in contact with paper strongly impregnated with tar, it would probably hurt the flavor, but wrapping tarred paper around a hive would not produce any such result.

Q. Would it pay to wrap hives in tar paper for spring protection? I see some favor it, while others do not. Why is there this wide difference?

A. There is a wide difference in climate. While it might not pay in the far South, it might pay well in the far North. Localities differ. Your apiary may be in an exposed place, where the wrapping mentioned may be of great service in warding off the chilly blasts of spring, while another apiary a mile away may be in such a warm corner that the wrapping is not so much needed.

Q. How can I protect my bees after putting them out of the cellar, with tar-paper. And how should it be put, when taken off, and what is accomplished for the benefit of the bees?

A. Whatever protection of the kind is given should be given just as soon as possible after the bees are taken out, as it is likely to be colder then than afterward. The time for taking off depends upon the weather; no harm to leave it on until fruit-bloom, or even till the first bloom is seen on clover. The advantage is that the bees are kept warmer, especially cold nights. Just how much that advantage is, it would be hard to say, no reports being yet given as to comparative results with and without protection.

Tartaric Acid.—Q. Will you tell us the result of your experience in mixing tartaric acid or other acids with sugar syrup for winter stores? I haven't had time to experiment in this, and this fall I followed Prof. Cook's advice, according to his book and put an even teaspoonful of tartaric acid into 15 pounds of syrup. I think this amount of acid is altogether too much for the amount of syrup. I believe much less acid in proportion would keep the syrup from crystallizing. The bees evidently do not like it.

A. Some years ago I had considerable experience in feeding up for several winters with tartaric acid in syrup. I used an even teaspoonful of acid for 20 pounds of sugar. I think it worked all right. How much acid that would be to a given weight of syrup depends on the strength of the syrup. For winter feeding I used five pounds of sugar to two of water, and that made a teaspoonful of acid to 28 pounds of syrup. Prof. Cook's teaspoonful of acid to 15 pounds of syrup looks just at first glance as if he made it about twice as strong with acid as I did. Whether he really did so depends upon the strength of the syrup. Referring to Cook's Manual, edition of 1902, page 266, where he mentions an even teaspoonful of acid to 15 pounds of syrup, it will be seen that he says: "We use equal parts of sugar and water." With the proportion of a teaspoonful to 20 pounds of sugar there would be a teaspoonful to 40 pounds of half-and-half syrup. That, against this 15 pounds of syrup, shows that he made it two and two-thirds times as strong as I did. Mine seemed to be strong enough. Of late years I have used no acid. If I fed at all I gave them half-and-half syrup in August or early September, and the bees made it all right without any acid. I cannot help thinking this is better than later feeding with acid. My feeders are becoming idle capital, as the pasturage has so changed that a fall flow may always be counted on.

Q. Yesterday (Dec. 30.) I was examining the colonies I am wintering in the cellar, and I found one that I concluded had starved. There was about 12 pounds of sugar in the combs candied hard. My record shows that on September 20 this colony had about 15 pounds of honey. I fed them 20 pounds of sugar syrup with one ounce of tartaric acid to each 10 pounds of sugar. I am afraid some of the rest of my bees will go the same route. One dislikes to lose them after feeding and giving them the best care one can. I have read of some who feed sugar syrup without using acid, and do not have any trouble, and it seems that there is little or no trouble where acid is used. Last fall some of my bees were carrying out candied sugar a week after I fed them. Is it possible that I have not been making my syrup right. The way I made it was to place a boiler of water on the stove and let it come to a boil, then add the acid and stir it well. I then set the boiler off the stove and stirred in the sugar. I used two pounds of sugar to one of water.

A. I must say there is something I don't understand about this sugar-acid business. There are those who, as you say, insist that no acid is needed, and they are very emphatic about it. A very few say that the sugar hardens in spite of the acid, and you are one of the unfortunate few. The time and manner of feeding may have something to do with it. If you feed as early as August or the first of September, and use more water than sugar, I don't believe acid is needed. Even if you feed heavier syrup, if you feed it slowly, there should be no trouble. But with late feeding of thick syrup, I should feel safer with the acid. I hardly see why you should fail; you used more acid than I ever used, and I never had any trouble, although I have had much experience. Still it is possible that the mode of proceeding may have something to do with it, and I'll tell you how I proceeded, when I fed late with heavy syrup: Water was put into a vessel on the stove, and when at or near the boiling point, sugar was slowly stirred in at the rate of 5 pounds of sugar to a quart of water. The stirring was continued until the sugar was dissolved, so that the sugar might not settle to the bottom and be burned. When the sugar was dissolved, an even teaspoonful of tartaric acid for every 20 pounds of sugar, previously dissolved in water, was stirred into the syrup, and it was taken from the fire. I would hardly suppose that your reversing the order would make any difference, still it might.

Toads.—Q. I have seen a frog or toad on the alighting-board of the hive, close to the entrance, late in the evening, just when the bees have clustered on the outside these warm, dry days and nights. I did not see the toad eat any bees at this time, but I wonder if he doesn't.

A. Yes, there has been a good deal of testimony that frogs and toads eat bees. Toads are such useful creatures in the garden that they may pay for eating a few bees by the number of injurious insects they destroy.

Q. How do you keep toads from eating bees?

A. Perhaps no better way than to raise hives so high that toads cannot reach the entrance.

Tongue of Bees.—Q. Are there long-tongued bees? I can hardly swallow that. I think that is only a selling point for those who have queens for sale. I have a few colonies, and I intend to get a few queens this summer, so if there are any with spliced tongues, that is the kind I am after.

A. There can be no sort of question that there is a decided difference in the length of bees' tongues. Able men on both sides of the ocean have settled it by actual measurement, and at least some of them have no possible interest in giving anything but the truth, unless they are bribed outright to lie—a thing that, for one, I cannot believe. But don't make the mistake of thinking that the bee with the longest tongue must necessarily be the best bee. Other things being equal, the bee with the longest tongue is the best bee. But other things are by no means always equal. The bees that still store the most honey are the best bees, whether their tongues are long or short. But when you succeed in getting the best storers, it is just possible that they may excel in tongue length.

Top-bars.—Q. What width of top-bars do you prefer, 1⅛ or 1 1-16? Is there any practical difference? I expect to make a number of hives, and want to get them right.

A. I am using 1⅛ inches with good results.

Q. Do you believe that a half-inch thick brood-frame top-bar will tend to prevent the bees building burr-comb on such frames, as well as the three-quarter inch top-bar? Which kind do you use?

A. I do not believe that the one-half inch will prevent burr-combs quite as well as the three-quarter. Mine are seven-eighths.

Trade Marks.—Q. How can a trade mark be obtained for labeling honey when working up a trade?

A. A trade mark is registered by the Government at Washington, D. C., in order to be able to protect it in case of infringement or copying. For the details to be followed in securing such registry, better consult a good lawyer.

Transferring From Box-Hives.—Q. When is the right time to transfer bees from box-hives to modern hives, and how?

A. Wait until the bees swarm (in your locality they are likely to swarm in May), then hive the swarm in an up-to-date hive and set it on the old stand, setting the old hive close beside it. A week later move the old hive to the opposite side of the swarm, and then two weeks later still, or three weeks from the time of swarming, when all the worker-brood will be hatched out, break up the old hive and add its bees to the swarm. Then you can melt up the old combs.

Q. How do you like putting a hive with one frame of brood over the colony to be transferred, and a queen-excluder between, when you catch the queen in the upper hive?

A. It will work all right. Here is something you may like better: Drum out all the bees, putting them in the new hive on the old stand, with a frame of brood, put on an excluder, and then the old hive. In 21 days the worker-brood will be gone from the old hive above and it can be taken away and the combs melted up.

Q. When transferring bees, will it hurt to have the old hive wrong side up until the brood hatches?

A. No.

Q. Will the following plan work well for transferring? Say I have five colonies in box-hives and wish to transfer, and I go to a hive to be transferred and smoke and drum out all the bees into the frame-hive except some to care for the brood that is in the hive at this time, which we suppose is in May or June; after which I set the old hive, for say five days, with its entrance closed over the frame-hive and with a wire-cloth between. After five days I replace the wire-cloth with a queen excluder, which I let stay for fifteen days, or one day before all the brood is hatched, then I put on an escape-board in its place; and when they have all gone down, take the old hive off, save all the good combs, and melt the others.

A. Yes; only it is hardly necessary to leave the wire-cloth as long as five days. Likely two days would be long enough—just long enough for the queen to get started laying below. Indeed, it might work all right to give the excluder at the start, and the less time the wire-cloth is left the better it will be for the brood above it.

Transferring From Movable-Frame Hives.—Q. Will you please tell me how to transfer bees from one hive to another? The hive they are in is poor, and I would like to get them into one with 9 frames.

A. Just exactly how it should be done, provided the bees are now in a frame hive, depends upon the size of the frame now in use compared with one to which you wish to transfer them. If the frame is shallower than the old one, you will cut down the comb so as to make it the right depth. If the new frame is deeper, put the comb in, and then cut pieces to wedge in on top, or which may be more easily managed, turn the comb so the present top and bottom may be at the sides, and then cut the comb just deep enough to go in the frame. Before taking out the first frame from the old hive, have an empty frame ready for it. Lay some strings on a table or something of the kind; on these strings lay the empty frame, then after putting the comb in, tie the strings. Of course, the strings must be laid in such a way that they will be distributed along the length of the frame, perhaps six or more of them, each string independent of the others. When you take out the first frame, brush the bees from it before cutting, and put it in its hive, after tying. Then move the old hive from the stand and put the new one in its place, and after that brush the bees into the new hive each time you take out another frame.

Q. Last fall I purchased three colonies of bees in home-made hives of the Langstroth pattern. I found that the frames were badly made, so that the combs were crooked—in fact, they zigzagged in every shape. I left them just as they were, fed the bees steadily all winter, and they are good and strong now; but will not get more honey than enough to feed themselves through the coming winter. I would like to get these bees out of the old hives. Would you advise transferring them at this time (August 3)?

A. Perhaps it may be as well to leave them as they are till next spring or swarming time. Still, it may be all right to transfer this fall, if you are sure of a good fall flow after transferring.

Traps.—Q. Is it necessary for a beginner to use a drone and queen trap?

A. No; and the advanced beekeeper gets along very well without it.

Q. If I use an Alley trap on a hive and the colony should swarm while I am away for a few days, will they stay around or near the hive any length of time, or will they leave if not hived the same day

A. The trap holds the queen, and when the swarm finds it has no queen it will return.

Trees.—Q. What kind of trees, other than fruit trees, can bees work on?

A. Oh, my! A whole lot more; more than I can tell you, and more than I know. A few are linden, locust, poplar, eucalyptus, maple, banana, black mangrove, wild cherry, etc.

T-Tins.—Q. What is a T-tin? I see in the American Bee Journal the way to make a T-super, but I do not understand what is meant by the T-tin.

A. A "T" super has no bottom, but to support the sections

has three tin supports running crosswise. Each of these is made of a piece of tin so folded that a cross section looks like a "T" upside down. You can buy T-tins of supply dealers for about a cent apiece, probably much cheaper than you can get a tinner to make them for. (See Super, "T.")

Q. How high up between the sections do your T-tins come? Don't you have to saw a place for them in the separators?

A. Some of my T-tins are three-eighths and some one-half inch high. Either does. No place is sawed in the separator, which rests directly on the T-tins. It would be bad to have the separator come down lower.

Q. I am informed that you use nothing but the T-tin in your comb-honey supers. It looks to me that they should be the best all around, but they say that the weight of honey will make the tins give or bend. What is your experience? The bees glue the wood-holders very tight in this locality. The wood separators are also troublesome.

A. Whoever they are that "say that the weight of honey will make the tins give or bend," it must be that they have never seen a T-tin, or else they are poor judges of the strength of ordinary tin. On the contrary, it would take a much greater weight to bend a T-tin than to bend any wooden support in use in supers. Remember that there are two thicknesses of tin standing one-half inch upright. I have had 3,000 T-tins in use for many years, and have never known one to be bent the slightest by the weight of honey. It would probably be all the same if the honey were five times as heavy.

Tupelo.—Q. From where does the tupelo honey come?

A. Tupelo (also called Gum) is a tree of the south. It is especially abundant in Florida, where it yields quantities of honey.

Uniting.—Q. Is it advisable to unite a strong colony with a weak one in July or August, or wait until spring?

A. If the one colony is quite weak, or if you are not anxious to save the queen, then you had better unite in the fall, since there is much danger that a weak colony will not winter through.

Q. What kind of perfume is sprinkled over bees when uniting two colonies to make them of the same odor?

A. I think peppermint has been used, and anise, cloves, or any other perfume might serve the same purpose.

Q. How will it do to use a fine spray of water to unite bees?

A. I don't believe it would do very well, but don't know.

Q. How will it do to sprinkle with flour when uniting bees?

A. It is practiced a good deal in England, but for some reason not much in this country. I think some have reported favorably, and some not.

Q. If, in the spring, one should have a number of weak colonies, could they be united with stronger ones and not have any fighting?

A. If you put two colonies together without any precaution, each one having its own queen, there is danger of fighting. A great many times I have safely united by taking one, two or three frames with adhering bees from one colony and simply placing beside the brood-nest in another hive. A safe way is to place one hive over the other with a common sheet of newspaper between. The bees will gnaw a hole in the paper and gradually unite peaceably.

Q. What is the best way to unite weak colonies? Shall I kill the queen, or will the bees do that?

A. The bees will destroy one of the queens, but it may be better for the beekeeper to attend to that job. There will be more peaceful uniting if one colony has been queenless for two or three days.

Q. (a) How do you work the newspaper plan for uniting two swarms?

(b) I have never seen it tried, but I presume one of the queens would have to be destroyed. What would be the proper way to manage it?

A. (a) It is a very simple matter. Take a sheet of common newspaper, spread it over the top-bars of the one hive. Of course the bottom-board will be under the lower hive, and the cover over the upper hive. There will be no sort of entrance or opening into the upper hive, and no bee can get out of it until the bees gnaw a hole through the paper. Within a day or so they will gnaw a hole in the paper big enough for a single bee to pass at a time and the bees will pass through and mingle so slowly and quietly that there will be no quarreling, gradually tearing away more and more of the paper. In a few days or a week you can put all the frames of brood in one story.

(b) If there is any choice of queens, kill the poorer, otherwise the bees will take care of the matter themselves. It is better if the lower hive remain on its old stand.

Q. When uniting, do you leave any combs in the hive above

the newspaper, or do you use an empty hive-body so the bees will go down quickly?

A. It doesn't matter whether empty combs are left in the upper story or not. The bees will unite just as quickly with or without them. Of course, after the bees have had time to unite, the two stories are reduced to one, the best combs of the two stories being selected to fill one story.

FIG. 25.—Two colonies united by the newspaper plan.

Q. I have just read about your way of uniting two colonies by putting paper between them. Did you ever try putting a queen-excluding honey-board between them?

A. Yes, I have united with an excluder between the two colonies. It is much the same as having nothing between the two stories. In some cases—perhaps in most cases—bees will unite peaceably when one hive is set directly over the other, with no excluder between. In some cases, of course, they would unite all right with an excluder. But too often it happens that if one hive is set over the other without any precaution, there will be a severe fight. In that case I doubt if the excluder would do any good. But the paper will. There is no possibility, with the paper, that one set of bees can fall upon the others en masse. It will take a bit of time for a hole to be made in the paper that shall let a bee through, and for some time there will be passage for only one bee at a time. In the meantime the two lots of bees are getting the same scent, ready to unite peaceably. At any rate, I've had one lot of bees killed when there was no paper between, and I'm not sure I ever had fighting when the paper was used.

Q. I have 50 colonies of bees in dovetailed hives, and want to keep but 25, spring count. How and when can I double them up? what should I do to keep the frames of larvæ and honey? And how may I keep the frames of comb during the winter?

A. Better wait till next spring to unite. If you unite this fall, there may be some casualties in winter, and you would not then have your 25 in spring. Even if you are sure of no winter losses in your mild climate, there are advantages in waiting till spring. There will be no question about care of combs through the winter, and by doubling 50 full colonies in the spring you are likely to have 25 stronger colonies than if the doubling were done in the fall; and 25 very strong colonies will take no more care than 25 weaker ones, and will store more surplus.

Uniting Swarms.—Q. Is it better to unite two swarms and make one big swarm out of two? And will I get more honey from one big swarm than I would get from two small ones?

A. You will be more likely to get more honey from uniting. In places where a strong flow continues very late, more honey may be had from the two kept separate.

Q. I would like to know the best way to double swarms up. If they both come out the same day, or a day or two apart, should I put the old colony on top of the new swarm? Should I take the bottom out of the top hive, or how can I get them together?

A. If they are only a day or two apart, the easiest way is to hive the second one in the same hive as the first, just as if the hive were empty.

Q. When hiving two swarms should I sprinkle or smoke them to make them go in the entrance?

A. If you dump them at the entrance they will enter of their own accord, without smoke or sprinkling.

Q. What is the best way to unite a swarm direct from the tree with a weak colony? I have just shaken it in front of the hive, but many of the bees were killed at the entrance of the hive.

A. Perhaps if you had shaken the bees off the combs at the entrance, so that the two lots of bees would run in together, there would have been less trouble. If one lot has an old queen and the other a virgin, they do not unite so well.

Value of Bees.—Q. How much is a colony of Italian bees in a modern hive worth, including super, sections, etc., in the spring, summer or fall?

A. There is no hard and fast rule about this, although the variation may not be so great as with box-hives. It may be from $. to $10 in the spring, and $2 or $3 less in the fall.

Q. (a) What could I afford to pay for ⏤ swarm of bees hanging where they clustered, if it was the first that issued from the hive?

(b) What would a good, strong colony be worth if it was in an old box-gum? How much if it was in a movable 8-frame hive?

A. (a) So much depends. One swarm may have two ar three times as many bees as another, even when both are prime swarms. In some places you might get a swarm for a dollar from someone who got little from bees and in another place an experienced beekeeper might not be willing to sell such a swarm for five times as much.

(b) Again there would be a great variation. A colony in a movable-frame hive might be worth in some places $7 or more, in other places $5 or less.

Ventilation.—Q. What is the best way to ventilate the hive?

A. It doesn't matter so much just how you ventilate, so you give ventilation enough. One way is to raise the hive by putting a block under each one of the four corners. I generally ventilate by having a very large entrance and an opening at the back end of the hive on top letting the super come far enough forward to make the opening.

Q. Does a hive need ventilation if in the shade? If so, would it need it when the temperature gets up to 90 degrees in the shade? How low can the temperature get before it needs shutting down?

A. Yes; I once had combs melt down in a hive so thoroughly shaded that the sun did not shine on it all day long; but there was a thicket on one side and a cornfield on the other, so that there was little chance for the air to stir. A colony must have ventilation to some extent always. At any time when the bees are busy gathering there should be sufficient ventilation so the bees will not hang out. An entrance equivalent to nine square inches is as little as should be allowed, and if that cannot be had otherwise, the hive should be blocked up. But 20 square inches of ventilating space is better than nine. There is no need to make any change when the temperature runs up to 90 degrees, nor when it runs down on cold nights.

Q. Do you use ventilation under supers, or open at the top through the summer?

A. Generally, with section-supers, I have ventilation at the back end, between the hive and lower super, and sometimes in the cover of the hive as well. In a cool time, however, it is better to have the ventilation closed, for sections at that part are not finished so soon.

Q. In your text-book you give a plan of ventilating the upper stories by shoving them forward and back, leaving a space at one end. Does the rain not get in through the space?

A. I suppose it does, but it never seems to do any harm, being at the end. At any rate, the harm is overbalanced by the good.

Q. How is the best way to ventilate hives in winter?

A. In the cellar it matters little how, provided there be enough ventilation and there is no danger of having too much. Formerly, with box-hives, a good plan was to turn the hive upside down, with no covering over it. That left it all open above and all closed below. Of course, no sort of hive ventilation will avail if the air in the cellar is impure. For outdoor wintering, the entrance may be three-eighths by six inches for a strong colony, and less for a weak one; besides this opening at the entrance, some cover with some sort of packing that allows a little air slowly to pass upward. Others leave the cover sealed down as the bees left it in summer and fall. But in this case the top must be warmly covered.

Q. What do you think of ventilation at the top of a hive in winter? Is it important, and, if so, would it not be proper to cut a 2-inch hole through a quilt and place the cloth cushions filled with cork chips on top of this? I use table oilcloth for quilts in summer and winter. Is there anything better?

A. There is a decided difference of opinion as to the matter of upward ventilation in winter, some reporting success with sealed covering, others reporting disaster. In either case it is important to have warm covering overhead for outdoor wintering. You are on the safe side not to have all sealed tight, and the plan you propose may work all right. I used oilcloth, same as you, for years, but for many years past have had no covering over brood-frames except the hive cover, and this method I like better. But it must be remembered that I winter in the cellar.

Vinegar.—Q. I am told that good vinegar can be made from honey or cappings. Will you give how much honey or cappings to each gallon of water and how to proceed to make it?

A. Use one to one and one-half pounds of honey to each gallon of water. Dissolve the honey and place in a barrel with the bung removed, so as to give as much air as possible. The warmer the place it is stored the better, as this will hasten fermentation. If you use capping washings for making vinegar, a good way to test if the water is sweet enough is by the use of an egg. If the egg comes to the surface of the liquid, then it is about right. To hasten fermentation, you may also add a little vinegar mother, if you have it, to your sweetened water. Full instructions may be found in most beebooks.

When you test honey water with an egg the egg should show only the size of a dime out of the water.

Waste Places.—Q. Does anyone know of something that could be sown in waste places where irrigating water runs, or where Bermuda grass now grows that would produce honey and also be good for the farmer? There are several places here where Bermuda grass grows, when it gets the waste water from the ranches.

A. Sweet clover has been very successful in such cases.

Water for Bees.—Q. What do the bees do with the water they get in the mud-holes?

A. The same as they do with water from any other source; they use it for drink and to thin their honey, for feeding the brood.

Q. We have an apiary where an irrigating ditch runs right along in front of the hives, but the bees go over to our neighbor's, about 80 rods away and get water from their watering trough, and they annoy them very much, as the stock can hardly get any water to drink on account of the bees. What could we do to help out our neighbor?

A. When the bees have formed the habit of going to a certain place for water, it is a very hard thing to get them to change to some other place. If the trough is not too large, it may be covered up by boards, sheets, or otherwise, opening it only at certain times in the day to let the stock drink. After a few days the bees will give it up. Possibly you may be able to make the place offensive for bees while still all right for the four-footers. Put carbolic acid or kerosene on the edges of the trough where the bees stand to get the water. Of course there is the danger that in doing this you will get some of the stuff in the water, so the stock will not drink it. As in so many other things, prevention is better than cure. In the spring, when bees first begin to get water, do all you can to prevent their getting a start in the wrong place, and to start them in the right place. In a sheltered place where the sun will keep it warm, put a tub or pail of water, throw over it some cork chips, such as grocers get as packing in kegs of grapes, and you will have a watering place where no bees will drown, and all you will need to do will be to fill up occasionally with water. Once started there, they will be likely to continue. One would be likely to think the bees would prefer the nearby irrigating ditch to the water trough farther away. But bees do not object to a considerable distance, and it is possible that the trough gives better footing for the bees, and that the water in it is warmer than in the ditch.

Water for Bees—Cork Chips.—Q. What size of cork chips do you use in water to keep bees from getting drowned? Also, about how thick is the layer of corks on top of the water? I am trying to get cork chips here. I can get granulated cork, of which I have samples, Nos. 2, 3 and 4. Watering bees in this locality is quite an item. My 75 colonies get away with as high as 60 gallons per day. I have to haul it all. I have been using a large trough filled with brickbats, but the brickbats take up almost all of the space. I also tried second-hand corks (cut them up), but in a few days the water would be foul; mostly wine corks. I am sending samples of cork chips. Should they be finer, or coarser, etc? (California.)

A. I don't believe it makes so very much difference as to the size of the cork chips, although I suppose the finest chips will lose their buoyancy soonest. Neither does it matter greatly as to the depth of the layer, only so it be not so thin that the bees will sink down into the water, nor so thick that they cannot reach the water. The chips I have been using are those which the grocers receive as packing in kegs of grapes that come in winter, or at least very late in the fall. The chips are of various diameters, from very fine ones, up to those that are one-eighth inch or more in diameter. A layer about three-quarters inch deep is first used, and more added later as they become soaked. The idea is to have enough chips so that the top surface will be a little out of the water. Although I never tried that size, I suspect that the coarsest you send (something like one-quarter inch in diameter) would be ideal.

Water, Bees Near.—Q. (a) Would bees be likely to do well near a large body of water, or would they be likely to fall into the lake?

(b) On which side of a lake would you prefer to keep—east or west side?

A. (a) The water is not likely to do any harm, only it is just so much surface without any pasturage, just like so much barren land. If the body of water was so narrow that the bees would cross it to get pasturage on the other side, a few bees might be beaten down in crossing by high winds.

(b) The side that had the best pasturage.

Wax (See Beeswax.)

Wax-Extractor.—Q. How can I make a solar wax-extractor without much expense? Does the solar wax-extractor take out all the wax, especially out of old combs?

A. Any kind of a shallow box, and of any size, covered with glass, so placed that the rays of the sun shall shine directly into it, will become hot enough on the inside to melt wax. A single pane of glass will do if large enough, or a common window-sash may be used. To hold the pieces of comb to be melted, have a plain sheet of tin, slanting 1 to 3 inches (according to the size of the box) from rear to front, so that the melted wax will run down into a vessel that you will place under to catch the wax. You may use a sheet of wire-cloth, so the wax will run through. This will work very nicely with cappings and burr-combs, but a good deal of wax will be left in old brood-combs. Especially will this be so if one brood-comb lies over another.

Weak Colonies.—Q. Would it be all right to put a new swarm in with a weak colony and thus make a strong one out of it?

A. Yes; but in thus uniting, the two queens should both be laying queens, or both virgin queens. If one has a laying queen and the other a virgin, they are likely to fight.

Q. I have two colonies of bees which I hived last May. One of them produced about 50 pounds of surplus honey, while the other produced only 5 pounds. What was the matter with the second one? Was it an unprolific queen, or not?

A. It may be that there was a difference in the strength of the two swarms at the time they were hived, and it must be remembered that a colony twice as strong as another will store a good deal more than twice as much surplus. The difference may have been in the character of the bees. Some bees are more industrious than others. There may have been other causes or a combination of causes.

Q. If you should find your bees weak in the spring in numbers, what would be the best way to strengthen them?

A. If I found a colony very weak in early spring, I wouldn't try to strengthen it. I would unite it with a stronger colony, or else I would wait till other colonies were so strong that they had at least six frames of brood each, and then I would swap its frame of brood for one from another colony. The frame taken from the weak colony would likely not be very well filled with sealed brood, and the one given should be well filled. Afterward more brood could be added, when the sealed brood had pretty well hatched out.

Q. In putting frames of capped brood into inferior colonies, is it not of importance to put in first one, or may a greater number be put in? I imagine that the surface of brood must be proportionate to the number of bees in a colony relatively weak.

A. You must use caution or you may have a lot of dead brood. If all the brood in the comb be sealed, and if it be old enough to be hatching out, then very little care is needed, for such advanced brood will keep up its own heat nearly as well as the mature bees. But you will seldom have such combs, and if there be considerable unsealed brood, or brood that has been sealed only a short time, then there must be enough bees in the hive to cover it well. One way to avoid chilling is to take the frame of brood with the adhering bees. Only if you add too many strange bees you may jeopardize the queen. Let the strange bees never be more than half as many as the bees already in the weak colony.

Q. As a rule, every beekeeper has some weaklings in his yard, I don't care how much attention he gives them. To strengthen them, what is your plan, to swap frames, or go to strong colonies, give them a good shaking and leave them with the queen and one frame of brood in the hive on the old stand, and put the rest of the brood under the weak colony? Very likely there would be queen-cells started.

A. Early in the season the former plan; at the approach of swarming, the latter.

Q. Did you ever practice the strengthening of a weak colony by reversing the hives, respectively, of a weak and a strong colony? As you seem to understand German, I will state that Berlepsch recommends this during a "volltracht," which, I suppose, means "full-flow." This looks very easy, only may be too late for securing full advantage from the strengthened colony.

A. I think I never tried strengthening in that way. There is danger of the death of the queen in the weaker colony unless in a time of full flow, and strengthening of weak colonies generally occurs before the heavy flow, or after it.

Q. What do you think of Mr. Alexander's spring management of weak colonies? What would you advise me to do to save the weak colonies in the spring?

A. Some of the things Mr. Alexander favored, it would be wild for others to follow, such as keeping so many colonies in one apiary, his special conditions favoring that; but as to the matter of weak colonies in spring, he has done the fraternity a real service. Care, however, must be taken. The first time I tried it the strong colony was at work inside of ten minutes fighting the weak one, and didn't stop till it made a finish. The colonies must be gently handled, so there will be no getting together till the upper colony has had time to get the scent of the lower, or else a wire-cloth must separate them for two or three days.

Weight of Bees.—Q. How many bees in a pound?

A\ A pound of bees may contain from 4,000 to 5,500.

Q. How much is the dovetailed hive, honey, comb, bees, etc., supposed to weigh just before putting them into the cellar?

A. I want my 8-frame hives to weigh at least 50 pounds. Ten-frame hives ought to weigh ten pounds more.

Willow-Herb.—Q. Is the willow-herb a cultivated plant, and would it pay to plant it for bees in Illinois?

A. Willow-herb is a wild plant found especially abundant in the burnt-over timber lands of the northern states. It is found in large quantities in Michigan and neighboring states, where it yields a very light honey. The honey is said to have a very fine flavor. Willow-herb is also known as fireweed.

Wind-Break.—Q. I have no wind-break or shade at home. Would it pay me to move my apiary 60 rods from home and have both?

A. It depends a little upon how much you care yourself for shade to work in, and how much, also, for the inconvenience of having them so much farther away. If you winter your bees in the cellar it would make no difference in wintering. It's a toss up which way you decide. Shade may be supplied by individual roofs on the hives.

Winter-Cases.—Q. With a regular brood-chamber and a winter-case made of seven-eighths inch lumber, is 1½ inches on sides and ends and 8 inches on top, enough for safe wintering?

A. Likely it will answer, although a greater space between

walls is generally used; say two or three inches. But doubling the space between walls will by no means double the protection.

Q. I am planning to build winter-cases 24x36 inches, each to hold two colonies. I also intend to leave the cases around the hives in the summer, as a protection from heat. Will the bees become confused and enter the wrong hive, or will queens returning from their mating trip be liable to enter the wrong hive?

A. I suppose your idea is that bees or queens may be confused by having the two entrances in what seems to them the same building. I don't think there will be any trouble in that way. I have used double hives with entrances not six inches apart, and I don't think there was any more trouble than with separate hives.

Winter Packing.—Q. Which is the best to put over the frames in winter, a solid board, a chaff cushion or a cloth, packing the super with leaves?

A. The cushion for outdoor wintering; in a cellar it matters little which, if the cellar is all right and the hive has a large entrance.

Q. I have always hesitated to remove the winter protection (chaff tray, etc.) in the early spring in order to examine colonies. As you advocate to take the bees out when maple is in bloom, would this also be a good signal to go by for removing the winter packing, or, if not, what would be?

·A. If you will look again you will see that it is the soft maple that usually gives the signal for taking bees out of a cellar. The hard maple blooms a little later. Taking bees out of a cellar is a different affair from taking away the wrappings of those that have been wintered outside. My bees have no wrappings after being brought out, but some think it pays to give them protection after that time. At any rate, if my bees were outdoors and well packed, I would hesitate about unpacking them at the time of maple bloom unless I thought there was danger of their being short of stores, and even then it might be worth while to return the packing until about the time of fruit bloom.

Winter Quarters, Removing From.—Q. How soon will it be safe to take out of their winter chaff-lined boxes and put on summer stands, bees that are in single-walled hives? (Iowa.)

A. If the bees are not to have their stands changed, and can have a flight without removing any packing, it is better for the bees to have the warmth of the packing until it is fairly hot weather, say about the last of May in your region.

Winter Stores.—Q. How much honey should I leave in each hive as a winter supply for the bees?

A. A store of 30 or 40 pounds is none too much for wintering outdoors, a stronger colony needing more than a weaker one, and for cellaring, 10 pounds less will do. Better five pounds too much that five ounces too little. The overplus will not be wasted.

Q. Are eight Langstroth frames full of honey enough to winter a strong colony of bees out-of-doors? I pack in leaves, three in a shed, six inches of space between each hive.

A. Yes, less than eight frames; six would do if well filled.

Q. Will the bees go through winter with as small an amount of honey as 15 or 20 pounds, when in the cellar?

A. In some cases they would, but it would not be safe to risk it.

Wintering.—Q. In preparing bees for winter, would it be best to leave the hive full of honey, or leave some empty combs for brood?

A. Don't you worry about room for brood. The best you can do at getting the brood-chamber filled with honey, no doubt there will be by spring plenty of room for brood, and the bees need no room for brood late in the fall. Some, however, think it better for them to have some empty cells to cluster on in winter, but they will have these emptied out in good time.

Q. Will it be safe to winter bees on combs with nearly all cells partly full of honey, but little or no capped honey?

A. Not very safe, but it might succeed.

Wintering in a Building Without Flight Opportunities.—Q. I wintered my bees in the granary last winter, and of 19 colonies only 8 lived through the winter. They seemed to be troubled with dysentery, and the stuff they passed was one-quarter of an inch thick on top of the frames. The last 4 hours of the bees' lives they seemed to pass nearly a teaspoonful, and all of very bad odor. This winter I left them on the summer stands with these results: From 18 colonies all but 3 died of the same disease. The hives are full of nice looking honey. Would it be all right to put a colony of bees in these same hives without removing the honey? I have an idea that the sudden change in temperature caused the hives to become damp, and thus the disease.

A. The likelihood is that the granary was too cold a place. A well ventilated cellar might give better results, being warmer. It is possible, also, that they were not packed warmly enough on the summer stands, especially on top. It is just possible, also, that the honey was at fault, but in that case it would likely be dark from honeydew. It will be all right to use these hives without removing the honey. Even if it should be honeydew, the bees can stand that all right when flying daily.

Wintering in a Building With Entrances Arranged for Flight.— Q. I have my bees in the attic, facing east, and it is so arranged that the temperature can be controlled during the winter months. During the most severe weather the past winter it has not been below 32 degrees, and never above 40 unless the weather out of doors was warm enough for them to have a flight. What would be the best temperature and cause them to consume the least amount of stores, with the hive-entrances open to the weather at all times as they are now?

A. About 50 degrees, but there is a good deal of variation in thermometers.

Wintering in Cellar.—Q. How many cubic feet per colony is required in cellar wintering? I am thinking of putting the bees indoors.

A. Something like ten, including passage way.

Q. In cellar-wintering, must it be dark in the cellar?
A. Yes, unless the bees keep perfectly quiet in the light. When first put in the cellar they don't seem to mind the light much, but do a great deal toward spring.

Q. In wintering bees in the cellar, do you leave the bottom-board off the hive for ventilation?
A. My bottom-boards are left on; but that still leaves abundant ventilation, for the space under bottom-boards is two inches, and the entrance is two inches deep and the whole width of the hive. If I had entrances not more than half an inch deep, I should want the hives blocked up or the bottom-boards taken away entirely.

Q. Do you put on anything to keep out the rats and mice, if such enemies should come along, or will the bees take care of their combs and honey in such a case themselves? I think Prof. Cook says that he leaves the bottom-board on and the entrance wide open.
A. I have done both ways. You may be sure the bees will not take care of themselves; rats and mice will make bad work if allowed undisputed possession. If you leave the hive-entrances open, in most cellars, you must keep up an unceasing warfare against rodents with traps and poison. You can bid defiance to the nuisances, however, by having the entrances closed with very coarse wire-cloth—three meshes to the inch. Even then you

will have some trouble, for field mice will have entered some of the hives before being brought into the cellar. It is better, however, to have a mouse confine its loving attention to one colony than to give it the free run of all.

Q. Do you recommend sealed covers for cellar-wintering, the cellar being damp?

A. I put my bees in the cellar with covers sealed down; but, they have entrances full width two inches'deep. With very small entrances there should be upward ventilation.

Q. Do bees always keep on humming in the hives all winter when in the cellar,. in which the temperature is 45 degrees? If not, please tell me the cause.

A. I believe some say their bees are found entirely quiet, but I think mine never are. A humming, more or less pronounced, may always be heard. They seem to go somewhat in waves, occasionally stirring up so as to make quite a little noise, but almost entirely quiet during the rest of the time. These periods of occasional waking up differ in different hives, so that when one stands to listen at the door of the bee-room there is a constant, gentle murmur, which I confess I rather enjoy hearing.

Q. There is a continual hum in the hives, sometimes sufficient to be heard across the cellar. Is this too much noise to call it good behavior and, if so, what is the cause and remedy? Temperature of cellar is 35 degrees.

A. Less noise would probably be better; but as cold as 35 degrees they will make a good deal of noise to keep themselves warm. Can't you warm up the cellar in some way? Hot stones or jugs of hot water tightly corked might do. Even if occasionally to 45 degrees or more, it would help.

Q. What difficulties may be expected from keeping bees in a damp cellar? How can these difficulties be overcome while the bees are in such a cellar?

A. Diarrhea is likely to result if the temperature is not sufficiently high. Bees have been reported as wintering in the best condition in a very wet cellar when the cellar was kept warm enough and supplied well with fresh air. Obviously the thing to do is to raise the temperature sufficiently, and to see that there is a sufficient change of air.

Q. What do you think of shutting bees in the cellar with wire-screen? I use a frame with wire-screen on both sides. I close up the entrance with two small nails and a strip of wood. I use another strip to hold all together, instead of staples.

A. I never tried it, but those who have tried it generally condemn the practice. I remember especially E. D. Godfrey, of Iowa, who, years ago, suffered loss by it. When the bees find themselves imprisoned, they make such a to-do as to stir up the whole colony. I have used wire-cloth at entrances in winter, but it was, of course, three meshes to the inch.

Q. Do bees in the cellar change the location of their cluster during the winter?

A. Bees do both ways, both in the cellar and out. Sometimes honey is carried from an outer comb, without changing the place of the cluster. Usually the cluster moves gradually backward or upward, as the bees eat their way into the full combs.

Q. When bees are fed in the cellar at a temperature of about 40 or 45 degrees, will the queen go to laying and hatching brood?

A. Hardly, unless the feeding be kept up regularly for some time.

Q. There is a whitish liquid running out of some of the hives in my cellar, and others have a dry substance like fine sawdust in front of the entrance. The hives from which the liquid comes seem to be wet inside and nasty. What is the cause of this? They have plenty of stores.

A. The sawdust-looking material is the gnawings from the cappings and other debris, and indicates nothing wrong. The liquid is the moisture from the vapors condensing in the hive. Your cellar is too cold, and hive-entrances hardly large enough.

Q. In American Bee Journal, R. H. Smith says the best temperature for wintering bees is 45 to 48 degrees above zero. If I remember rightly, all our best authorities agree on 42 to 45 degrees for most successful wintering in cellars. I have one Standard barometer and three Fahrenheit thermometers. One of the latter is filled with quicksilver or mercury, and the others with colored fluids. I have all these in my cellar, and the variation from the one that shows the highest to the one that shows the lowest, is 10 degrees. Upon which can I depend for the desired 42 to 45 degrees which is necessary for successful wintering of bees, as claimed by our best authorities?

A. So you're up against that mixed matter of temperature in cellar. Latest investigations seem to show that the right temperature is about fifty or fifty-five degrees. But, as you have found out, thermometers vary. You will also probably find that cellars vary, perhaps on account of the difference in dryness, perhaps for some other reason, so that if the same thermometer is used in two cellars, it may need to be higher in one than the other. I don't know which of your thermometers is best, and it doesn't make very much difference, although on general principles it's better to have it correct. But here is what you're to do: Take whichever thermometer you think best, and keep close watch until you find at what degree your bees are quietest, then keep your cellar as near that temperature as you can, whether it be 42, 45, 55, or something else. The idea is to find at what temperature your bees are most quiet by your thermometer, in your cellar, no matter what authorities may say.

Q. When in the spring is the right time to take bees out of the cellar? I can't find it in any of the beebooks. (Missouri.)

A. It isn't an easy thing to say when is the right time to take bees out of the cellar, and I'd give a pretty penny to anyone who could tell me with certainty the best time to take mine out this spring. There has been as much as a month difference between the earliest and the latest of my taking out, there being that difference in seasons. There must be more or less guessing about it so long as one never knows in advance just what the weather is going to be. So long as they are in good condition in the cellar, and there is nothing for them to do outdoors, there's no hurry about taking them out. If you will watch the blooming of red maples, willows, or other trees upon which they work in your neighborhood, you will generally find it best to take them out at the time of such bloom, but not even then if the weather appears unfavorable. So far south as you are—in Missouri, 39 degrees—are you sure it is advisable to cellar bees at all?

Q. We have had trouble about our bees rushing out when taken from the caves and cellars, all getting mixed, apparently, and when returning fill some of the hives full of bees and leave others badly weakened, so as to make it detrimental to the depopulated hives. Would a wet rag stuffed in the entrance be good, leaving only room for a few bees to pass in and out at once, or would simply closing the entrance almost entirely answer?

A. I confess to you that there are things connected with your question that I don't understand. Every year, for many years, I have taken out my bees with a rush, taking them out so that all could have a flight that first day. Others say that when they do that, the bees swarm out and make lots of trouble, but I have never had any serious trouble. Some say to take out a few each day. That would hardly work here, for when it comes time to take bees out of the cellar there may not be two days in succession fit to take them out. Indeed, the two good days may be several days apart. Possibly one reason for the difference lies in the condition of the bees. The night before mine are taken out, doors and windows are open to the widest, and all night long they have fresh air. So, when they are taken out they do not feel the change of air, and often they do not fly out of the hive at all for some minutes after being put on the stands. If taken out of the close air of the cellar when they are uneasy, they may get so excited that they will swarm out.

A big rag made very wet is one of the best things to lay against the entrance when you want to keep bees in temporarily, but I doubt that it would help any in the case under considera- tion. I would rather have the bees so quiet that there is no need to fasten them in. Try giving them a tremendous airing the night before taking out. Contract the entrance immediately on setting the hive on its stand.

Wintering Out-of-Doors.—Q. How is this for wintering bees: In the late fall, after the honey-flow is over, place a piece of bur- lap over the brood-frames and place a super filled with dry leaves on top of that? (Tennessee.)

A. For your locality it would be hard to find anything better.

Q. What is the best plan to adopt in an effort to winter bees on the summer stands?

A. I would rather trust a single-walled hive in a protected place, sheltered by buildings or trees, than a double-walled hive fully exposed to the sweep of winds. If no other protection is at hand, go back to that of our grandfathers. Make a shed, under which the hives stand in a row, only a little higher than the hives, closed on all sides but the side of the entrances, and then pack straw in all the vacant space inside the shed. Even cornstalks piled about a hive, wigwam shape, produced quite good results with one man not five miles from us.

Some have an outer case allowing a packing of leaves, planer shavings or other loose material about the hive to the extent of three or four inches on all sides and probably six inches on top. Others omit the packing-case and hold leaves in place by a frame of coarse netting.

Q. On account of my apiary being some distance from my cel- lar, I am thinking strongly of wintering outdoors. My bottom- boards are reversible with the deep side two inches. My hives are 10-frame dovetailed. (a) Should I use the deep side for win- ter? (b) How much of the entrance should be closed?

(c) Would it be safe to wrap the hives with extra heavy tarred felt, with no other protection? (Wisconsin.)

A. (a) Yes. (b) The equivalent of two or three square inches will probably do well where you are, the latter for a very strong colony.

(c) Probably, but something depends upon the exposure of the situation. If exposed to the full force of the winds, it will be hard to wrap the hive warm enough, but in a situation well sheltered from the winds there will be little trouble.

Q. Do those who winter bees on the summer stands need to scrape out the dead bees?

A. It is better, of course, to have the dead bees cleaned out. With the usual shallow entrance, and shallow space under bottom-bars, it may be absolutely necessary; for the entrance, other-. wise, may become entirely clogged with dead bees. With a deeper space under the bottom-bars, and entrance at the upper part of the space, cleaning out the dead bees is not so important.

Wintering on Super-Combs.—Q. (a) I ran short of beehives and have about five swarms which I hived in shallow supers used for extracted honey. They have ten shallow frames. I wish to know what you would do with them, unite them with other swarms, or let them winter in these supers, and in the spring put them in the regular hive?

(b) How would you go about it to put them in the regular 10-frame hive?

A. (a) Unless they are weak and you want to unite them with other weaklings, better let them winter as they are.

(b) In the spring set the shallow story over the regular hive, which should have frames filled with foundation. When brood appears in the lower story, put a queen-excluder between the two stories, making sure that the queen is in the lower story. Eight or ten days later kill any queen-cells that may be in the upper story.

Wintering in Two-Story Hives.—Q. A Tennessee beekeeper writes that he wintered most of his 180 colonies in two-story hives and he never had such strong colonies; some had 15 brood-frames. Why wouldn't that be the best way to do every winter?

A. With very strong colonies the plan is excellent.

Wiring Frames.—Q. Is it necessary to wire shallow extracting-frames when medium brood foundation is used?

A. Hardly, especially if care be taken at the first extracting, emptying one side only partly, reversing and extracting the other side, and again reversing to empty the first side.

Q. If I wire the shallow 6-inch frames, can I use extra thin surplus foundation in them? How many wires ought I to put in, and where should they be?

A. I fear you could not use extra-thin without four or five wires. You could probably use thin surplus foundation with two horizontal wires, one two inches below the top-bar and the other one and one-half to two inches lower.

Q. Is it necessary for frames to be wired?

A. Not absolutely necessary, but better, to have the combs strengthened by being supported by wires or foundation splints.

Q. Is vertical wiring as good as horizontal? If not, why not?

A. That depends. If top and bottom-bars are sufficiently rigid, vertical wiring is as good or better. With vertical wiring, the wire must be drawn tightly, and unless a bar of some kind is in the center to hold top and bottom apart, the bottom-bar will be curved upward, and if the top-bar be not pretty thick it will sag.

Q. What do you think of using wire from baled hay or straw in place of your wooden splints in brood-frames?

A. Such heavy wire would be objectionable. Only very fine wire is used in wiring frames.

Worms in Bees.—Q. Sometimes when I take off the lid there is a worm crawling on the underside of the lid about an inch long and one-fourth inch thick, gray color. Can that be some of the larvæ that got out of some cell, or is it some other prowling stock?

A. That worm is not an escaped larva from one of the cells of brood, but "prowling stock" of another sort. It is the larva of the wax-worm, which destroys combs when they are not properly protected by the bees. These prowlers are not worth minding in strong colonies, or those of good Italian stock, but when a queenless colony is on hand, especially a weak black one, these moth larvæ finish up, like a lot of crows about a carrion. (See Beemoth.)

Worker-Bees.—Q. Can workers lay?

A. Not as a rule; but when a colony has been queenless a long time they may undertake the business, and then we have the pest called drone-laying workers. (See Laying Workers.)

Q. How many days from the time the worker hatches until it goes to gather honey?

A. It is generally understood that a worker goes afield when 16 days old. But the wise little creatures know how to adapt themselves to circumstances without following any rigid rule. One time I had a valuable queen to introduce. Over a strong colony I put an empty hive, with wire-cloth between the two stories. In the empty hive I put frames of sealed brood with young bees just ready to emerge, but not a bee. I put in the queen and closed up tight, so no bee could get in or out. Five days later I gave a very small entrance, and the bees flew. A little later some of the bees returned with loads of pollen. Those babies, only 5 days old, were doing work that under ordinary circumstances they would not have done until three times as old. So in the economy of the hive while bees generally go afield when 16 days old, the likelihood is that they accommodate themselves to circumstances. If conditions are such that there is an unusual need of nurses, some of the bees may not go afield until considerably more than 16 days old, and vice versa.

Yellow Jackets.—Q. I notice yellow jackets entering some of my hives. Do you think they are doing mischief? The bees do not seem to notice them.

A. They're probably after honey, and are not likely to get off scot-free.

Yields of Honey.—Q. Is 100 pounds per colony an average, or toward the maximum yield?

A. Hardly one or the other. One-third of that amount is nearer the average yield of comb honey, and half of that for extracted, while a maximum annual average might go to 150 or more. It must not be forgotten that the yield sometimes is less than nothing; that is, no surplus is taken and the bees have to be fed to keep them alive.

Author of "Stray Straws" from December, 1890, to December, 1919.

DR. C. C. MILLER, the beloved friend of all beekeepers, died in his ninetieth year, Sept. 4, 1920, after a final illness of five days. There was no dimming of his person-ality during the later years of his life. Until the very last he remained at his best, ever alert, genial, full of enthusiasm, always radiating a great-hearted love that embraced all nature and all mankind. Dr. Miller's life was one of the richest blessings of the beekeeping world, and his writings will be a most prized inheritance for years to come.

Early Life.

Dr. Miller was born of English and German parentage in a country home at Ligonier, Pa., June 10, 1831, and here he spent his boyhood days, enjoying life to the full. The country surrounding his home was beautiful, and awakened in him that great love of nature that was so characteristic thruout his life. At the age of ten years he lost his father, whom he greatly loved and revered. In his writings he has characterized him as "most lovable in character," and has stated that thruout his life he has been influenced by the desire to be as good a man as his father.

His Education.

By working in a country store two years at $24 and $50 per year Dr. Miller obtained enough to go to the village academy. He was then obliged to teach before taking up his college work at Jefferson College, Cannonsburg, and Union College, Schenectady, N. Y. By rigid economy, boarding himself at 35 cents a week and doing any honest work from ornamental penmanship to peddling from house to house, he completed his course, taking at graduation the highest honor, Phi Beta Kappa.

A Physician.

After one term of teaching in Geneseo Academy, N. Y., he studied medicine at Johnstown, Pa., and attended lectures in Michigan University. He received his M. D. degree, and for a year practiced medicine in Earlville and Marengo, Ill.; but he was not happy in the work. His health was not rugged enough to stand the strain, and he was so vitally concerned that each patient should show immediate improvement under his care that the responsibility of his profession proved too great, and he was obliged to take up other employment.

Music and Teaching.

At the age of 26 Dr. Miller married Mrs. Helen M. White. A few years were spent in teaching vocal and instrumental music, and a few years as principal in the Marengo public schools.

In 1870 and 1871 he traveled for the music house of Root & Cady. In 1872 he spent six months as official agent in starting the first of the May musical festivals under Theodore Thomas at Cincinnati. The three following years he worked for the Mason & Hamlin Organ Co. at Chicago, his wife and little boy leaving the farm and spending their winters with him. During the summer months, when they were not with him, however, visions of the country continually haunted him, making the city appear desolate indeed; and so in 1876, in spite of an offer of $2500 and expenses, he left the city and took a school at Marengo at $1200.

His Beginning with Bees.

Altho Dr. Miller when a boy had taken a little interest in a colony of bees that his father kept in a barrel, still he had given bees but little thought until 1861, when his wife captured a runaway swarm and hived it in a barrel. This colony the first year stored 93 pounds besides teaching Dr. Miller a great deal about bees. Eight years later he saw a copy of the American Bee Journal, and among other interesting writers he found the name of A. I. Root, whom he visited at Medina the following year. For the first nine years but little increase was made; but in 1876, when he gave up his city work and returned to the country, he had 99 colonies. From this time on he made beekeeping his business.

In the spring of 1880 his wife died, and in the fall of 1881 he married Miss Sidney Jane Wilson. Her sister, Miss Emma M. Wilson, was his main assistant in the apiary after that time up to his death.

Some remarkably good honey crops were secured by Dr. Miller. The best record was an average of 266.74 sections from 72 colonies, and his best colony that year produced 402 sections.

His Writings.

It is doubtful whether any one else was ever as well informed in beekeeping literature as was Dr. Miller. He always attempted to read all the journals on beekeeping, even those in German and French. His own experience, thus backed by the experience of others, made him an exceptional writer. Moreover, his wit, tact, and unfailing good humor endeared him to the hearts of his readers.

Dr. Miller was always at his very best when assailing another's position on any given subject or when defending his own. For this reason he was prevailed upon in 1890 to contribute the department "Stray Straws" for Gleanings in Bee Culture. This department was continued without interruption until last November, when failing health made less work imperative. Since 1894 he has conducted "Dr. Miller's Answers" in the American Bee Journal. The separate articles, also, that he contributed to the different journals from time to time were always valuable and right to the point. His "Fifty Years Among the Bees" has been an exceedingly popular book.

Love of Nature and God.

As Dr. Miller said, he might easily have amassed more money in some other line of work; but in so doing he certainly could not have taken the enjoyment that he had in his quiet country home among his flowers and his bees.

Dr. Miller was a life-long member of the Presbyterian Church. To him the spiritual life was all very real. He not only believed in it but he lived it, as was testified by every act of his splendid life and by every feature of his wonderfully expressive face.

www.ingramcontent.com/pod-product-compliance
Lightning Source LLC
Chambersburg PA
CBHW051334200326
41519CB00026B/7416

9 781908 904379